Understanding the Mathematical Way of Thinking – The Registers of Semiotic Representations

Raymond Duval

Understanding the Mathematical Way of Thinking – The Registers of Semiotic Representations

Edited by Tânia M. M. Campos

Springer

Raymond Duval
Université du Littoral Côte d'Opale
Dunkerque, France

Translated from the Portuguese language edition: "Ver e ensinar a matemática de outra forma: entrar no modo matemático de pensar: os registros de representações semióticas" by Ruth Maria Vidotti Kakogiannos, Copyright (c) 2011 Proem Editora Ltda

ISBN 978-3-319-56909-3 ISBN 978-3-319-56910-9 (eBook)
DOI 10.1007/978-3-319-56910-9

Library of Congress Control Number: 2017939868

Printed on acid-free paper

This Springer imprint is published by Springer Nature
The registered company is Springer International Publishing AG
The registered company address is: Gewerbestrasse 11, 6330 Cham, Switzerland

Foreword

After devoting his time to the study of problems related to learning the concept of infinity in mathematics and to the distinction between demonstration and argumentation, in the early years of the 1990s, Raymond Duval surprised the world of research in the didactics of mathematics with an intense succession of studies into the decisive importance of semiotics in the activity of learning mathematical concepts (Duval 1993).

We had sporadic meetings during that time, particularly during the ICMI Study Conference, on "Perspectives on the Teaching of Geometry for the 21st Century", in Catania in 1995 and, months later, in July 1996 in Seville, at the ICME 8 where I had the opportunity of collaborating with Raymond Duval in the topic group I led (as chief organizer) on "Learning infinity: Infinite processes throughout the curriculum," although by that time his research efforts were focused on semiotics (Duval 1995).

His approach, while drawing from the conceptual bases of Frege, De Saussure, and Peirce, is decidedly revolutionary. The sense he gives to his research is strictly related to the learning of mathematics, and his now world-renowned phrase "There is no noesis without semiosis" has been pronounced by all of us and is one of the phrases most cited in our research world.

Another universal contribution is the idea of the "cognitive learning paradox," in which it is stated that it is paradoxical that the student constructs a mathematical object, by having only semiotic representations of O and not knowledge of O. Moreover, semiotic representations are the only means by which teaching might show O. This is a position with thousands of years of history, since the time of Agustín de Tagaste, who said:

> Cum enim mihi signum datur, si nescientem me invenit cuius rei signum sit, docere me nihil potest: si vero scientem, quid disco per signum?» (For when I am shown a sign, it cannot teach me anything if it finds me ignorant of the reality for which the sign stands; but if it finds me acquainted with the reality, what do I learn from the sign?) (*De magistro*, 10, 115).

In a recent work (D'Amore et al. 2015), we show precisely how this brilliant idea of Raymond results from the itinerary of philosophical-semiotic positions that began in ancient Greece: I personally communicated this fact during an international conference in Santa Marta, and Raymond expressed great interest in this analysis.

During my visits to Lille, where he lives, Raymond was always generous in every way. So much so that now I can say with certainty that his principal works reflect, in my opinion, only a small part of his real knowledge. He knows how to disseminate this knowledge in many forms during research seminars, in supervising doctoral students, in master's courses, in meetings with teachers, in colloquia with students, and beyond. Those who know him and who hear him speaking at such events can experience how the didactic transposition is valid at every level of schooling and is indeed an exciting subject.

Personally speaking, I think I only definitely understood the differences between Vygotsky and Piaget, thanks to his explanations, one day, in Lille, at his house, while trying, without much success, to prepare a lunch—even after having read all that has been written about this subject. This was also the case as regards my understanding of and the position of De Saussure.

He quotes frequently, and also in this book on pages 27 and 28, the American artist Joseph Kosuth, famous for being one of the first artists of the so-called analytical line (D'Amore 2015a). Among so many works, Kosuth is famous for "One and Three Chairs," made from 1966 onwards in so many versions and currently housed in the best museums in the world. In this, he exposes a real chair of various materials, a photograph of the chair, and a definition of the word "chair" taken from a dictionary.

It is a work of great importance in the world of art history and also interesting to the field of semiotics. I am happy to inform Raymond in these lines that the French artist Bernar Venet, in 1996, also exhibited a work entitled "Tube," which is based on the same idea: the object (a tube) and an axonometric representation (D'Amore 2015b), another indication of this time of operation. I think that for a study of the importance of semiotics in art, it is necessary to start with the Belgian René Magritte (D'Amore 2010).

In this book, Raymond explores his semiotic world as a giant, as only he can do—simply and briefly, but, at the same time, profoundly, learnedly, and concretely.

It proposes the relationship between representation and knowledge, thanks to a revolution within the semiotics (Duval 2006) and based on the role of representation of the knowledge of an object, in general, and of a mathematical object, in particular (Duval 2009a). He also presents the difference, from the cognitive point of view, between sign and representation (Duval 2009b).

Semiotics thus presents itself as a new scheme of knowledge analysis, which requires an in-depth discussion of the three fundamental models of the analysis of signs, each with its contributions and its limits: De Saussure's model (structural analysis of semiotic systems), Peirce's model (classification of the various types of representation), and the model of Frege (the semiotic process that produces new knowledge). (See Raymond Duval's three articles in Duval and Saenz-Ludlow 2016, and my specific comments in the same text).

A central problem has to do with the relationship between mathematical activities and semiotic transformations. In the process of accessing mathematical objects, two epistemological situations are evident, one irreducible to the other. A test (called an opposition) is used with a material object: how to recognize the same object in different representations and how to create correspondence between objects or between representations.

Transformations of semiotic representations are placed by Raymond at the center of mathematical activity. Here, the author is led by convincing examples of great epistemological and didactic force already published in other texts on geometric figures and natural numbers (Duval and Saenz-Ludlow 2016). From these reflections, he concludes that a cognitive analysis of mathematical activity and the functioning of the thought in mathematics are necessary.

We then proceed to the registers of semiotic representation and to the analysis of the cognitive functioning of thought in mathematics: the important difference between codes and registers, the analysis of the types of discursive operations and cognitive functions of natural languages, the relations between thought and language, and the characterization of a semiotic representation register.

This is followed by an analysis of one of the bastions of research in didactics of mathematics: visualization. How do we see a figure? How do we see the transformations of a figure? How does all this work in the didactics of geometry? The examples presented here are portentous and precious.

A complete chapter is devoted to registers. Here, we find considerations about the unity of meaning in mathematics relative to the content of a representation, how mathematical activities vary according to the registers that are put into play, functional variations of the phenomenological modes of production in relation to registers, and how to achieve meaningful and useful analysis of classroom activities.

The power of this work, as with others of the same author, is that the relation between studies, apparently only theoretical, and classroom life is total. He has frequented many classrooms, and I have read his narratives of the activities developed with students. In fact, the theoretical flights that sometimes seem to leave behind the concrete classroom situations are always occasions to recapture living spaces, where the interests are closer to those of the students and the teachers, in order to reach the highest peaks.

These living spaces have no age limits. Some of his considerations are appropriate for children in the first grades of schooling, while others seem to have no reference to age at all. Indeed, I am always surprised to think of the formidable epistemological analysis of his work, which treats the classroom as an experimental environment for the construction of mathematical objects.

His very personal way of seeing, in the specific sense of "seeing," figures is a remarkable contribution to epistemological research in mathematics; I find confirmation in others of his studies (Duval 2016).

The step taken toward the world of art is most impressive, since it provides the scholar with two semiotic fields, the artistic and the mathematical, considered not at all interchangeable in the common sense.

The study of this brief book will undoubtedly be of benefit to those who carry out research in the field of mathematics education, as well as to those who work daily in the world of school and wish to eliminate the gap between the mathematical objects that we intend our students to build cognitively and those that, in reality, the student constructs. This is treated as, as if this was, an epistemological problem, yes, but a problem of a concrete didactical nature.

Raymond himself suggests four different modes which might be adopted while reading this book: a linear reading, the most straightforward and diffuse, from beginning to end; a synoptic transversal reading of recurrent themes; a practical reading, looking for themes and explicit didactic suggestions; or reading it as if it were a cartoon or a book for children, looking only at the figures and following them in their evolution. I personally followed the first modality, and, being fortunate to know in detail these subjects, I appreciated the structure and choice. I then tried to follow the figures, as a child would. I must admit that a certain amount of courage is necessary, but it is undoubtedly an interesting modality of reading.

To the reader, now, is the choice.

Bruno D'Amore

Bibliographic References

D'Amore B (2010) Figurative arts and mathematics: pipes, horses and meanings. En: Capecchi V, Buscema M, Contucci P, D'Amore B (Compiladores) (2010) Applications of mathematics in models, artificial neural networks and arts. Mathematics and society. Springer, Dordrecht, pp. 491–504

D'Amore B (2015a) Arte e matematica. Metafore, analogie, rappresentazioni, identità fra due mondi possibili. Dedalo, Bari (Italia)

D'Amore (2015b) Bernar Venet. Elogio del processo razionale. Nuova Meta. 37:2–13. www.rivistaartenuovameta.it

D'Amore B, Fandiño Pinilla MI (Compiladores) (2015) Didáctica de la matemática. Una mirada epistemológica y empírica. [Autores: Guy Brousseau, John Alexander Alba, Luis Carlos Arboleda, Ferdinando Arzarello, Giorgio Bolondi, Ricardo Cantoral, Bruno D'Amore, Raymond Duval, Martha Isabel Fandiño Pinilla, Vicenç Font, Athanasios Gagatsis, Juan Díaz Godino, Salvador Llinares]. Actas del congreso internacional homónimo, septiembre 2015. Ediciones Universidad De La Sabana, Chía (Colombia)

D'Amore B, Fandiño Pinilla MI, Iori M, Matteuzzi M (2015) Análisis de los antecedentes histórico-filosóficos de la "paradoja cognitiva de Duval". Revista Latinoamericana de Investigación en Matemática Educativa 18(2):177–212. http://www.clame.org.mx/relime.htm. doi: 10.12802/relime.13.1822

Duval R (1993) Registres de représentations sémiotiques et fonctionnement cognitif de la pensée. Annales de Didactique et de Science Cognitives 5(1):37–65

Duval R (1995) Sémiosis et pensée humaine. Registres sémiotiques et apprentissages intellectuels. Peter Lang, Berne (Suiza)

Duval R (1998a) Signe et objet (I): Trois grandes étapes dans la problématique des rapports entre représentations et objet. Annales de didactique et de sciences cognitives 6(1):139–163

Duval R (1998b) Signe et objet (II): Questions relatives à l'analyse de la connaissance. Annales de didactique et de sciences cognitives 6(1):165–196

Duval R (2006) Quelle sémiotique pour l'analyse de l'activité et des productions mathématiques? En: Radford L, D'Amore B (Compiladores), Semiotics, culture and mathematical thinking [Special Issue]. Revista Latinoamericana de Investigación en Matemática Educativa 9(1):45–81. Obtenido de http://www.clame.org.mx/relime.htm

Duval R (2009a) "Objet": un mot pour quatre ordres de réalité irréductibles? En Baillé J (Compilador), Du mot au concept: Objet. 9–108. PUG, Grenoble (Francia)

Duval R (2009b) Sémiosis, pensée humaine et activité mathématique. Amazônia Revista de educação em ciências e matemáticas 6(11):126–143. Obtenido de http://www.periodicos.ufpa.br/index.php/revistaamazonia/index

Duval R (2012) Quelles théories et quelles méthodes pour les recherches sur l'enseignement des mathématiques? Práxis educativa 7(2):305–330. doi: 10.5212/PraxEduc.v.7i2.0001

Duval R (2016) Voir et créer dans l'art et en géométrie: proximités et divergences. En: Iori M. (compilador) (2016) La matematica e la sua didattica. Mathematics and Mathematics Education. In occasion of the 65 years of Bruno D'Amore. Proceedings of International Conference, October 8, 2016. Department of Mathematics, University of Bologna. Preface of Bruno D'Amore. Pitagora, Bologna (Italia). ISBN: 88–371–1927-5. Descarcable gratuitamente de los sitios: http://www.dm.unibo.it/rsddm, http://www.incontriconlamatematica.org, http://www.incontriconlamatematica.net. 213–220

Duval R, Saenz Ludlow A (2016) *Comprensión y aprendizaje en matemáticas: perspectivas semióticas seleccionadas*. [Prólogo de Bruno D'Amore. Comentarios a los artículos de Bruno D'Amore y de Carlos Eduardo Vasco Uribe]. Universidad Distrital Francisco José de Caldas, Bogotá (Colombia)

Preface

I could not have written this book if it was not for the invitation of Professor Tania Maria Campos Mendonça to present a seminar for the doctoral students of the mathematics education graduate program of the Universidade Bandeirante de São Paulo (UNIBAN). The seminar focused on the teaching of algebra to all elementary school students, analyzing its difficulties and exploring other introduction strategies. Therefore, the main purpose was not to present the registers of semiotic representation, but to understand how they allow analyzing the cognitive processes that we should encourage and develop in students, so they understand and use the basic tools that algebra provides. Thus, the presentations and materials for the working sessions were prepared for this purpose.

Very quickly, the request of the doctoral students and the program's professors, who participated in the seminar, was for the representation registers. A large proportion of the participants knew the distinction between the kinds of transformations of semiotic representations: conversions and treatments. Some participants had seen the relevancy of this distinction to identify the students' understanding difficulties and to analyze the cognitive processes that underlie understanding and non-understanding phenomena during the mathematics learning process. But that raised several important questions for the seminar participants, such as when and how to use such a distinction in the research work related to mathematics education. So, the seminar focused on the following issues:

- How to situate the representation registers in relation to other semiotic "theories"?
- Why use a semio-cognitive analysis of mathematical activity for teaching mathematics?
- How to distinguish the different kinds of registers?
- How to organize tasks and learning activities that take into account the registers as variables?
- How to analyze students' productions in terms of registers?

The theoretical aspect was less important than the analysis tool and work method. The questions and discussions that arose, in a way, determined the plan of this book.

This book is based on the analyses developed in *Sémiosis et Pensée Humaine* (1995), but the main ideas are presented from another perspective.

In *Sémiosis et Pensée Humaine*, analyses focused on the cognitive processes of mathematical reasoning, in which the natural language is explicitly used, even when symbolic formulations are used. The purpose was to analyze the specific cognitive process of reasoning that underlies proofs in geometry. And that led us to consider that the language and its use are not only constituted by the words used but also by the discursive operations employed when speaking and reasoning whether in a deductive, argumentative, or purely semantic way. It is only when these discursive operations are taken into account that language appears as a true representation register and the relationship between language and thought can be systematically examined.

In this book, on the contrary, we are more interested in the other registers, especially those that allow visualizing in mathematics. Figures in geometry, graphs, and the various types of tables used in statistics or in other areas raise at least as many difficulties as reasoning in geometry. But we pay less attention to them because we believe that in this case, since we "see," the recognition cognitive processes would be the same as in any iconic representation, such as images, diagrams, maps, and photographs. However, this is clearly wrong.

The various kinds of representation used to visualize in mathematics become a source of misunderstanding, all the more important that they have an increasing place in the teaching of mathematics for two reasons. First, the emphasis is placed on practical activities that involve using a lot of varied, iconic representations, such as figures, "curves," and tables, among others. Second, the use of a computer for everything that concerns mathematical visualization, both in geometry and in analysis, and geometrical or graphical software opens considerable possibilities of creation and visual exploration. But does software suffice to develop in the students the ability to anticipate the different possible transformations of a given figure into others completely different? Does it make students aware of the one-to-one mapping between graphic visual values and the terms of the equations they represent?

Another issue has also become more explicit after the publication of *Semiosis et Pensée Humaine.* It is about the particular epistemological status of mathematical knowledge in relation to the other domains of scientific knowledge. Certainly, didactics is concerned about epistemology, but it is only an intra-mathematical epistemology, in which we follow, historically, the formation steps of the concepts or mathematical objects, each with its own motivations and obstacles. This intra-mathematical epistemology cannot be considered an epistemology of scientific knowledge; insofar it does not take into account the formation of concepts in chemistry, geology, botany, or paleontology. The ways to get access to these objects of knowledge and the kinds of proving are not at all the same as in mathematics. This fact raises a serious question regarding the learning process of students who must switch between maths and other scientific disciplines and, therefore, different ways of working on the same day or the same week. Does teaching really help them to realize how the way of thinking and working in mathematics is different from that in other sciences?

In fact, the specific understanding problems that students face when learning mathematics are rooted in the particular epistemological status of mathematical knowledge, not only on questions of the pedagogical organization of activities. Indeed, the way we get access to mathematical objects is radically different from the way we do for the objects of other scientific disciplines. This is the crucial point for learning mathematics, and it is also the first challenge in the teaching of mathematics. It is necessary to develop a kind of cognitive funtioning wholly different from the one mobilized in the practice of the other sciences so that students understand how learning mathematics contributes to their global intellectual development. This goes against the general direction of most research on mathematics teaching, where it is assumed that the model to acquire and build knowledge would be the same for all areas of knowledge, mathematical and non-mathematical. In order to see and to teach mathematics differently, we must, instead, be aware of the specific cognitive processes that mathematical thinking requires and develop them with the students, even if by doing this, teachers have the impression "of not teaching (momentarily) math"!

And algebra? In the seminar, we then turned to the issue of learning to use equations as a problem-solving tool. This acquisition is generally the main objective of teaching algebra in elementary school. But where to start? By introducing letters and literal calculation to solve numerical problems? But this produces insurmountable obstacles for many students. Cognitive analysis leads to another approach. First, it highlights the specific discursive designation operation of any object that putting data into equation requires as opposed to the designation operations in the practice of ordinary language. Second, it emphasizes the diversity of conversions underlying the problem statements. Therefore, other initial goals and tasks that consider the registers used for producing a problem statement are required. Thus, about these two crucial points, the introduction of letters and problem-solving, it is necessary to:

- Dissociate the introduction of letters from any problem-solving activity.
- Associate *the first use of the letters to the functional designation of numbers.*
- To teach how to make up problems that can be solved mathematically so that students become able to solve any problem.
- Practice all the various conversions of representation that can be involved in any verbal description of non-mathematical situations. Conversions required to solve a problem are not the same when starting from the description of a real situation or the equations that allow solving the problem of this situation.

What matters in introducing algebra at middle school is that students develop an awareness of specific cognitive operations required to put data into equations and to be able to recognize whenever equations can be used to solve real problems.

This analysis was developed in the theoretical framework of registers of semiotic representation. And it provides an application of this theory to an area of mathematics, where, unlike geometry, cognitive activity seemed reduced to pure mathematical operations! In a way, the two parts of the seminar form a whole. Nevertheless, we felt it important to publish them in two volumes.

This book can be read or used in four different ways.

We can do a more or less complete *linear reading* as with any other work. Just following the text, issues such as the justifications of the concepts and explanations of the distinctions necessary to analyze the cognitive functioning of mathematical thinking are introduced gradually. This reading allows us to understand the coherence of the entire line of thinking and to learn the internal relationships that form the theoretical framework of analysis. However, this reading retains only an overall impression that is often of little use, unless we do several local readings.

A cross and synoptic reading is also possible. We do not start from the text any longer, but from the terms that crystallize the strengths of the theoretical framework or analysis tool. Therefore, we enter into the text via the index terms. But, here, we must avoid the trap of the contents of dictionaries, hit list, and keywords. These indices isolate one term from another as if their meanings were not linked to each other. We should not consider a single term, but several and, if possible, a group of terms. The terms and pairs of terms are regrouped under a word, which condense them, to facilitate the cross reading. We only grasp the reality and the complexity of semiotic phenomena when we understand all the distinctions that characterize them.

This point is essential because it touches on a very frequent confusion between two types of conceptualization. One seeks primarily to have definitions, and even definitions presented with the help of some uppercase or lowercase letters, which we then apply to what we observe or even allow us to build a general model. In didactic research, these definitions relate almost always to crystallizer words and are limited to them. The other type describes the phenomena so that we will be able to identify the variables and the underlying processes that make their complexity. But we do not define a priori the relevant variables. To identify them, it is necessary to isolate them from systematic observations. And is it necessary to remember that the relevant variables we seek to identify are about the understanding/misunderstanding processes in mathematics learning? In this way of conceptualizing, the condensed words do not matter as much as the set of variations and distinctions they regroup in a descriptive network.

A practical reading is another possibility; it starts from examples and shows the way to analyze them. To facilitate this reading, we made a list of representations of objects and mathematical operations that are briefly presented here, but whose explanation can be found in other publications. The choice of the examples highlighted during the seminar has been guided by our experience of working with teachers and sometimes with teachers from different disciplines looking for an interdisciplinary collaboration. However, in this practical reading, we must never limit ourselves to one example, but consider at least two very different examples. A theoretical approach should allow researchers and teachers to analyze the understanding process in algebra as well as in geometry or calculus. In all these areas, the cognitive way of thinking and working is the same, and it has nothing in common with the one mobilized in geology, chemistry, or botany.

Finally, *we can look at it as a cartoon*. In this case, we start from all the figures that are part of the text. So we can start by looking at the two pictures that open Chaps. 1 and 2, respectively. They give an insight of the epistemological and cognitive issues raised by the analysis of knowledge and its formation: the necessary distinction between representation and object, the various kinds of representation, recognition of the represented objects, the conditions to get access to the objects, etc. Similarly, we can start by looking at eight small sequences of figures in Chap. 3 to become aware of the heterogeneity of figural operations underlying visualization in geometry. Here, we have a first spectral decomposition of the different ways of "seeing" required in any geometric activity. How do we expect to advance in the research about the teaching of geometry if we do not take into account all the figural operations that constitute the specific cognitive variables for geometric visualization? How can teachers understand the difficulties of their students if they are not aware of this cognitive complexity and, instead, adhere to a syncretic notion of an overall figure opposite to any drawing always particular?

There is obviously no difference among these different types of reading. We start as we want. But all of them are needed to see how the complexity of the studied phenomena leads to the development of a set of distinctions corresponding to experimentally isolable factors and to master the cognitive way of analyzing mathematical activity.

Raymond Duval

Acknowledgments

This work was supported by the Foreign Professor Visiting Program (PVE) of the Coordenação de Aperfeiçoamento de Pessoal de Nível Superior (Capes)– Process 0368-11/20.

Contents

List of Figures

Chapter 1
Representation and Knowledge: The Semiotic Revolution

Learning mathematics raises problems of understanding that are not observed in other areas of knowledge. There are, in particular, two radically different kinds of difficulties. Over a period of several weeks, or sometimes in one classroom session, there are the local difficulties related to the introduction of a new concept or a new procedure. Over the period of a year, and especially over a curriculum cycle, there are the global and recurrent difficulties related to solving a problem, reasoning, geometric visualization, graphic visualization, the inability to transfer and apply the acquired knowledge to new situations and reality. We can, of course, observe these global and recurrent difficulties during a lesson or sequence of activities, but they seem to be confused with local activities. That is why, in many works and models, one tries to explain these global and recurrent difficulties as if they were the result of local difficulties related to misunderstanding of some concepts.

To understand the underlying causes of these global and recurrent difficulties is not enough to observe what the students do and explain, what they succeed or fail, in the tasks and problems we give them. We must first ask ourselves *"what is the mathematical knowledge?"* and *"why is it different from other types of knowledge?"* These questions are both epistemological and cognitive, and these two aspects cannot be separated. The analysis of knowledge should not only consider the nature of objects studied, but also the way the objects are presented to us and how we can access them on our own. This question "how can we access on our own" is obviously essential for the formation and mathematics learning. It is at the heart of what we call understanding, and in mathematics, it cannot be reduced to proofs and justification. This, in fact, is the issue of the cognitive processes that are required in any mathematical thinking.

All models that refer to either the formation of knowledge or the cognitive functioning of thought must respond to three types of questions. Do we have a direct and immediate access to objects (for which we often use the general and plurivocal word "intuition")? What are the systems, structures, which are required to get the objects, either directly or by a sequence of conscious or non-conscious processes? Finally, what is the nature of the cognitive relationship between these processes and

R. Duval, *Understanding the Mathematical Way of Thinking – The Registers of Semiotic Representations*, DOI 10.1007/978-3-319-56910-9_1

knowledge objects? These three issues are articulated in what we call a knowledge analysis scheme. For beyond the terminology variation of each model, we always find the crucial distinction between the representations and the objects themselves.

Regarding the analysis of what is mathematical knowledge, the crux lies whether SEMIOSIS is considered or not. A true semiotic revolution began with the emergence of algebra and analysis and acquired an unprecedented scale in the nineteenth and twentieth centuries. In this sense, we are led to distinguish two different schemes for the analysis of knowledge; those that take into account the central role of semiosis in mathematical thinking and those that do not.

To have an insight into all these issues, and the complexity of notions they involve, the fastest way is to see how they emerged with the development of mathematics, and then with the one of the other sciences. We will see how the relationship between the notions of representation and sign has been reversed. But the sign notion is deceptively simple. This, indeed, is the reason why the founders of semiotics did not assign the same definitions to it. The comparison of their analyses will allow us to bring out the important notion of *semiotic representation*, and to not reduce the role of signs in cognitive functioning of thought to a simple coding of either information or concepts.

1.1 The Fundamental Epistemological Distinction and the First Knowledge Analysis Scheme

The first knowledge analysis scheme was based on the epistemological opposition between the representation of an object and the represented object. Knowledge begins when we no longer mistake the representation of an object for the object itself. In other words, knowledge begins with the awareness of this question:

(Q.1) When we think to be in the presence of an object, is it the object itself or a representation?

Although the immediate perceptual obviousness is always experienced as intuition or direct apprehension of things, it does not escape that question, as showed by the picture below shows it, does not result from a montage (Fig. 1.1).

If the reflecting surface of the building was entirely of glass, the resemblance would have been such that we would ask ourselves whether the two objects are similar or even we would have been unable to tell on which side the real object is! It is this kind of image, the reflection of trees and sky on the water that Plato (Republic VI, 509d–510b) used to exemplify the fundamental epistemological split that builds knowledge.

Paradoxically, this kind of example teaches us more about the representation than what constitutes the object itself. It allows to highlight the two characteristics of every representation.

(P.1) There are always many possible representations for the same object.

Fig. 1.1 Paulista Avenue in São Paulo, Brazil

The multiplication of possible representations of the same object has two origins. First, there is the indefinite proliferation of same kind of representations. Thus, the images produced by the reflection of light on a surface vary with the incident angle, the shape of the surface, etc. But, there is also a *diversity of the kinds of representations*. All representations are of the same nature. Plato (*Republic* X, 596–598) had already referred to the representations that he called "imitation", that is, intentional reproductions of a model such as in the paintings. There are other kinds of representations such as those obtained with scientific instruments (astronomical telescope, microscope, oscilloscope ...), those constructed in mathematics (geometric figures, graphics, equations ...). And we can add to this limited, non-exhaustive list, all the representations that are not directly observable, such as visual memories of new faces seen, the dreams, etc. This particular multiplication of representation kinds leads to characterizing also the representations according to its origin, that is, according to the system that allows producing them and that determines *the capacities and limits of the representations produced.*

(P.2) The diversity of representations of the same object originates from the variety of physical or semiotic systems that can produce representations.

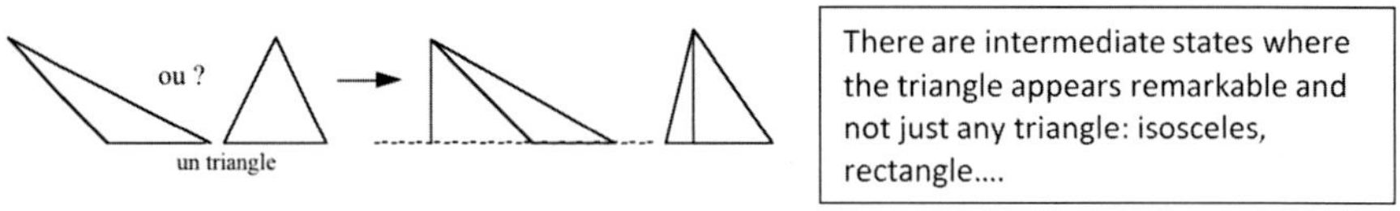

Fig. 1.2 Variability of figures and invariance of the geometric object represented

These two characteristics of the representation show an intrinsic difference between the representation and the object itself: the variability of one and the invariance of the other. Unlike the object, its representations change at the same time according to the considered viewpoints and the systems used to get a representation of it. The object appears as the invariant of all the possible variations of its representations. Thus, in geometry, the triangle is none of the particular figures we use to represent it, because a figure constructed with an obtuse angle, has two of its heights external and not internal, to the triangle (Fig. 1.2).

Finally, we must take note of the fact that variations of representations depending on the point of view are often confused with the variations of phenomena. And yet they are very different since the former not associated with time, while the latter cannot be dissociated from time. It is here that we find the ambiguity of Platonic opposition between the 'visible' world and the world of 'ideas', and also the ambiguity of Husserl's phenomenological analysis of perception. They merge phenomenal variability with representational variability.

1.1.1 The Cognitive Issue of Access to the Objects Themselves and Role of Representations

The analysis of knowledge focuses on how we get access to the objects themselves. The opposition between a direct access mode, described as 'intuition', and indirect modes that are based on formation processes and involve systems that make up the different mental faculties has first imposed itself as obvious. This knowledge analysis scheme based on the fundamental epistemological distinction aims to answer the three following questions (Fig. 1.3).

The questions (1) and (1a) are closely related. We cannot answer one without determining the response to the other. So, question (1) is answered in three steps. First, we have a direct and immediate access to all the objects within the multisensory perceptual field, which constitutes the surrounding reality. Therefore, the immediate perception is the starting point of all knowledge, even if sometimes it gives rise to illusions in which the representation is mistaken for the object itself. On the other hand, for the objects that are out of this field, but always subject to possible immediate perception (e.g., when moving or during a trip), we must rely either on our memory or the descriptions from others. Finally, the issue of accessing objects falling outside this scope remains open. The answers varied between a direct access

THE EPISTEMOLOGICAL DISTINCTION	**OBJECT**	⇄	**REPRESENTATION**
THE COGNITIVE MODES OF ACCESSING THE OBJECTS	(1) *Do we have access to the objects themselves, independently of representations?*	(2) *What is the nature of their relationship?*	(1a) *Which are the representation systems that allow access to the object?*

Fig. 1.3 The first knowledge analysis scheme

(Plato) and an indirect access involving a new type of representation: the purely mental or intentional "concepts".[1]

Question (2) is essential to the formation of concepts. It has, in fact, two aspects. The first is the way its formation is linked to the representations coming from perception. And this gave rise to different kinds of cognitive models: an ascending abstraction process (analogous to the passage of the particular to the general), a reverse schematization process (application of the first notions to sense experience that is analoguous to the one of words to things), and as a process of invariance awareness.[2] The second aspect concerns the nature of the relationship between the representations and the represented object. In this first knowledge analysis scheme, *the relationship has always been defined in terms of causality*. The debate lies on whether the object is the cause of representations or, conversely, is the spirit, as it is seen in the different knowledge analyses proposed by Descartes and Kant.[3] This aspect is obviously important because *the relationship of the signs to what they mean cannot be defined in terms of causality, but in terms of reference*, as we shall see.

1.1.2 Sign and Representation: The Cognitive Divide

The notion of sign emerged when knowledge analysis focused on how a verbal expression means something; or rather *how a verbal expression communicates something to someone*.

To understand how the sign notion emerged, let us consider the verbal expression in its most common practice, the speech. The speech is the intentional production from somebody. In the speech, the verbal expression always has two completely different faces depending on whether you are the speaker or listener. The listener

[1] Duval, R. (1998). Signe et objet (II): questions relatives à l'analyse de la connaissance. *Annales de didactique et de sciences cognitives*, n. 6, p. 165–196.

[2] Duval, R. (2007). La conversion des représentations: un des deux processus fondamentaux de la pensée. In: *Du mot au concept: conversion*. Baillé Jacques (Ed.). Grenoble: Presses Universitaires de Grenoble, p. 9–45.

[3] Duval, R. (1998). Signe et objet (I): trois grandes étapes dans la problématique des rapports entre représentation et objet. *Annales de didactique et de sciences cognitives*, n. 6, p. 139–163.

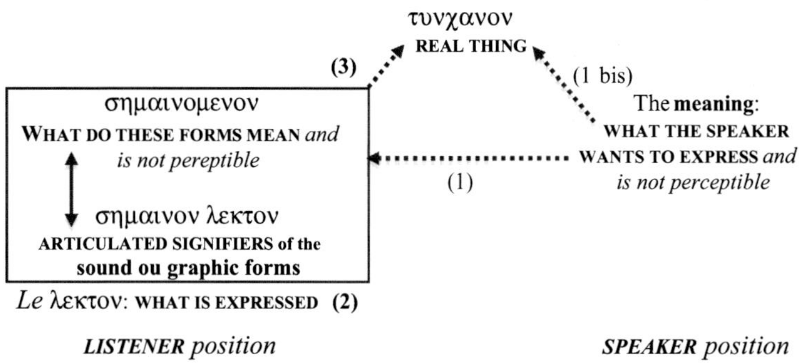

Fig. 1.4 The two faces of verbal expression and the notion of a sign

has only the pronounced words he heard (Fig. 1.4, (2)) to understand what the speaker wanted to say (1a). For the speaker, in contrast, the situation is quite different. The speaker's attention is entirely focused on what he wants to say, and little or not at all on what he says. Not only the speaker pays little attention to the accuracy of what he says (λεκτον), but also it does not distinguish the things he intentionally aims at (1) and the meaning of the words he uses (σημαινομενον). In other words, there is an important gap between the two sides, and it is not the same face of verbal expression that appears first, it depends on whether we are the speaker or listener. This duality of verbal expressions allows to understand the well-known distinction used to characterize a sign.

(P.3) The distinction between the "signifier" and the "signified" refers to the duality of all verbal expression as intentional production.

Thus, the Stoics, who had developed the first analysis of signs, had to consider verbal expression as the result of an intentional production. Naming it "σημαινον (the signifiers of what is said)", they placed themselves as listeners, which allowed them to make two other distinctions, as shown Fig. 1.4.

First, an internal distinction on the verbal expression itself and to which they employed two non-conjugated forms of the verb σημαινειν ("mean"). Of course, *the signifier and the signified cannot be separated,* no more than the verse and reverse of a coin to resume the comparison of Saussure. There is no word without meaning. The second distinction is external to the own verbal expression. It allows defining the second characteristic of the signs (Fig. 1.4; (3)).

(P.4) The relationship between the signs and the things they mean is a RELATIONSHIP OF REFERENCE and not a causal relationship.

This characteristic is essential since it allows seeing where the signs differ from the representations in general. The relationship with the objects themselves is never a causal relationship, but exclusively a reference situation.

The reference relationship is because the verbal expression is always intentional. In other words, it is not only the meaning of words in the language that allows the listener or the reader to understand the relationship of verbal expression with the object it describes or sets, but it is the intentional use that the speaker makes of the words. Now, this intentional use is realized by the designation operations. So, when we talk about an element of a geometric figure "the center of a circle," or "the middle of a segment that is a diameter" or "let O be the point ...", these verbal expressions refer to an element of the figure, because they result from a discursive intentional operation of designation.

(P.5) A reference relationship results from a DISCURSIVE INTENTIONAL OPERATION OF DESIGNATION.

Now, let us check the verbal expression not as the result of an intentional production, but the act of its production, i.e., from the speaker position. Can we, then, speak of signs? Husserl favored this position in his analysis of signs and their role not only in communication, but especially in thought. Starting from the duality of all verbal expression in communication, he separated in the words what is phonetically separable, but not semantically: the signifier and the signified. And he concluded that we would not need words when we think « in our mind », since the words serve only as indices about what the speaker means for the listener.[4] In any case, one thing is certain. The words to the one who speaks and when he speaks are transparent to what he means; and what he means is like a direct access to the object of which he speaks (Fig. 1.4, (1) and (1bis)). The *signs are transparent* to the one who is in the speaker position.[5] Can we conclude from this transparency of signs that signs would not be used in the thinking act, not only as Husserl did, but as many continue to do so in mathematics education when they speak of "mental representations"?

We can summarize all these remarks about the representation phenomena, the signs and their roles in the first models of analysis of knowledge, in the following four conclusions.

First, the representations are epistemologically ambivalent because on one hand they should never be confused with the objects themselves, but, on the other hand, they are always necessary to have access to the objects. They stand for the objects or "evoke" them when objects are not immediately accessible.

The signs are representations, and they should never be mistaken for the objects to which they refer to. That is why signs are defined by their common characteristic with the representations, as seen in this classic definition: "The sign is a thing (res) that, besides the form perceived by the senses, brings to mind the idea of something else ... ".[6]

However, the signs are radically different from the representations by the fact their relationship to the objects themselves is not causal but referential. And *this is what makes the difference between the cognitive power of all semiotic productions*

[4] Husserl, E. (1901). *Recherches logiques* 2, deuxième partie. Trad. Elie, Kelkel, Schérer. Paris: PUF, 1972.

[5] We exclude, of course, situations when the speaker prepared previously his speech by preliminary work. In this case, he repeats or reproduces what he had already expressed, read or heard!

[6] Augustin (1997) p. 136.

(statements, symbolic expressions, graphs plot), which are intentionally constructed by using semiotic systems, and cognitive power of representations, which are automatically produced by physical or neuron systems, as we shall see in the next two chapters.

Finally, the signs have no role in the first knowledge analysis scheme. All cognitive models drawn from this first analysis scheme privileges the representations produced causally. Even the abstraction and schematization processes, which fall within the set of faculties or systems that constitute the knowing subject (the sense, the memory, the imagination, the understanding, the reason), fall for the most part outside any monitored production of the subject.

1.2 The Semiotic Revolution: Towards a New Knowledge Analysis Scheme

The emergence of algebra as a symbolism that will be "the language of calculations" according to the famous formula of Condillac marks a new stage in the development of mathematics. It takes place in a historically short period. If we take as reference points the historically significant works, it happened over a short period of a century or a century and a half.[7] The introduction of letters replacing quantities and numbers raises for the first time *the issue of the role of signs in mathematical thinking, as seen in the work of Leibniz, a contemporary of Wallis.* And with that issue comes up a new opposition regarding the knowledge based on symbols and knowledge based on intuition. This opposition is irreducible to the classical cognitive oppositions between sensible and intelligible knowledge or between material realities and concepts, or between concrete and abstract! This opposition will even tend to supplant them, at least in the field of mathematics and science.

Many historians consider that with Descartes, algebra is set up. But, to stop there would be to see only one side of the emergence of signs and symbols that are specific to mathematics. The other side is the construction of graphical representations of geometric figures and, more particularly, all the proper conics (ellipses, parabolas, hyperbolas) by Descartes (1637). The decisive innovation was the coding rule, which associates two numbers to one point on a plane defined by two graduated and oriented axes or more. This rule allows going from equations to the graphical

[7]Cardan (1545). *Ars magna*. Viète (1591). *In artem analyticam Isagoge*. Descartes (1637). *La géométrie*. Wallis (1685), Algebra. Al Khwarismi, IX century, is the precursor, or the initiator, with the delayed translation of his works to the XIII century, of the constitution of algebra as a tool for problem solving. This period must be distinguished from the following two: the one when "the algebra joined analysis" and the letters became variables, and the other "when the calculations relate to the increasingly diverse objects: congruence, vectors, and arrays, set", led to the "arrival of the structures (Duperret, 1999, p. 30–34). In other words, after this first period, algebra has been developed as a function of the extension of calculation objects. This last period is different from the two preceding due to the fact that the objects are no longer what the letters refer to, but the "operations" we perform about what the letters can "designate".

construction of all the geometrical shapes we want (lines, curves, surfaces, etc.). *Hence the "objective" articulation*, seemingly simple, which enables the mind to work simultaneously on two kinds of semiotic representations quite different, as if it *were the same representational functioning*! But what is common between the syntactic organization of the equation terms and the visual characteristics of lines, curves or surfaces on a plane? Nothing. The respective units of meaning are heterogeneous and do not require the same kind of apprehension. In one case, the organization is based on terms whose linear apprehension is successive. On the other, the organization is based on some figural qualities of lines, which are perceptually recognized at first glance! How can we recognize whether they represent the same object or not, and whether two visually similar graphs represent the same mathematical object or not? Here we are faced with the requirement of not confusing the representations and the represented object (above P1). This requirement is no more an obstacle only when the two semiotic systems of representation *work cognitively in synergy*.

This semiotic extension of the field of representation paved the way for the introduction of new notations and enabled the invention of infinitesimal calculus a few decades later (Newton, 1669; Leibniz, 1672). It also thrown new light on the nature of this activity and mathematical knowledge. Leibniz was the first to be fully aware. To him, mathematical knowledge was a *"symbolic knowledge"*, even if he considered it still "blind" regarding the "intuition" of the objects themselves, although intuition is strictly limited to a few objects, as Descartes had shown, insisting on the limited capacity of apprehension of the human spirit (1628).[8] Leibniz observed that symbolic knowledge can overcome this finitude of intuition by providing unlimited possibilities for indirect access to mathematical objects.

This semiotic extension of representations led to a revolution in the approach to the question of signs and their role in the cognitive functioning of thought, even beyond the fields of algebra and analysis. This will translate into two radical changes in the first knowledge analysis scheme. On the one hand, the central role is given to the signs in the working of mathematical thinking. On the other, there is the need for an *epistemological distinction between objects according to their cognitive modes of access*. This leads to the following knowledge analysis scheme (Fig. 1.5).

Note that the top line of the scheme preserves unchanged the epistemological opposition in which the first knowledge analysis scheme was based on (Fig. 1.3). In the first scheme, the signs were not radically different from the representations, and their use was subordinated to the represented or assigned objects. In other words, *their cognitive function was reduced to that of any representation:* to stand for the objects themselves when these are not immediately accessible, that is to allow evoking the objects in their absence. Signs were distinguished from representations only by *their communication role*.

In this new scheme, the signs form a specific circuit with its autonomy. They are not just a external supplementary step following a sequence of representation pro-

[8] Descartes (1972). *Règles pour la direction de l'esprit.* In: *Œuvres philosophiques*. Ed. Alquié. Paris: Garnier, 1972.

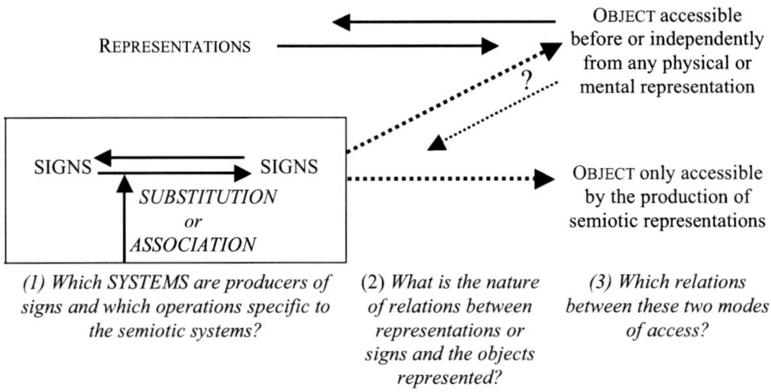

Fig. 1.5 The new knowledge analysis scheme

ductions leading from the perception to concepts. This central role of signs results from elimination of their subordination of their only use to communicate. This elimination opens the possibility of semiotic transformations, exploiting all possible combinations, which are no longer limited only to what is empirically real or verifiable. The signs then fill a *cognitive and epistemological treatment function* to produce new information or to establish new knowledge.

(P.6) The main property of signs is that they can BE SUBSTITUTED ONE FOR ANOTHER INDEPENDENTLY OF the objects they may evoke.

The question (1a) of the first analytical model of knowledge should, therefore, be framed again into two questions (Fig. 1.5, (1) and (2)). The driving force of unlimited substitution of signs one for another (the two opposite horizontal arrows in the rectangle of Fig. 1.5) is not the knowledge of represented objects, but the operations specific to the semiotic system that has been used to produce representations of the objects (the two arrows starting from the lower level). And, as we shall see, these two questions are crucial to examine the cognitive processes underlying the mathematical way of thinking.

This first radical change leads to distinguish epistemologically at least two kinds of objects *depending on how we can access them* (Fig. 1.5).

On the one hand, there are those which are accessible to perception. This access may be direct and multisensory for everything, which lies in our perceptual field, but it can also be indirect and monossensorial with the use of instruments that change the perception scale, as the telescope, microscope, etc. Let us remember that the semiotic revolution was contemporary of the scientific revolution that began with Galileo, i.e, with, on the one hand his first astronomical telescope and, on the other hand, his first substitutions of measured values for the qualitative sensory data in experiments on falling bodies. In these situations of instrumental accessibility, the relationship between the representations produced and the objects themselves is a cause-effect relationship from what is beyond perception to our eyes through eyepieces, screens, etc. The diversity of representations results from the kinds of instruments used.

On the other hand, there are the objects that are not at all accessible, neither concretely nor instrumentally, and for which we should intentionally use sign systems. The relationship between the representations thus produced and the represented object is no longer a causal relationship, but a relationship of reference.

The question then lies in knowing to which of these two categories the mathematical objects belong.

1.3 The Three Models of Sign Analysis That Have Founded Semiotics: Contributions and Limits

It was not until the late nineteenth century that the signs were systematically studied. Certainly, the sign notion had already been brought out by the Stoics and, in the eighteenth century, Condillac had compared the natural language with mathematical symbols from the point of view of the economy and power for calculations. But, the study of signs itself truly began with the first analysis models of the diversity of signs and their role in the scientific activity and communication.

Three models appeared almost simultaneously, and independently. Two are explicitly associated with the foundation of "semiotics" as a discipline: Peirce's model between the years 1890–1910[9] in the United States, and Saussure's model, with the publication in 1916 of *Cours de linguistique générale* in Geneva, about the structural analysis of language. The Frege's model, mainly with the two seminal papers published in 1892 and 1894,[10] and made famous by the Russell's criticism in 1905 [Russell, B. (1905)]. All subsequent work on semiotics originated from the contributions of these three authors.

These three models have nothing in common. Neither the same definition of sign nor the same analysis criteria to distinguish the types nor the same description of cognitive functioning that they enable. The reasons are, firstly, the area of knowledge selected to study signs and their use. For Peirce, these were the sciences, in general, without distinction, and logic. For Saussure, who had worked on the development of Indo-European languages, it was linguistics. For Frege, these disciplines were mathematics and, more precisely, Analysis and arithmetic. Can these models be relevantly generalized to areas of knowledge other that the one they selected? Secondly, the guiding questions that determined their research were not the same (Fig. 1.6)!

Therefore, we see the issue of the use of any semiotic approach to understand the learning process in mathematics. Which model is most appropriate to analyze the productions of the students and the activities proposed for the purpose of acquiring knowledge? Is it the model of Peirce, Saussure, and Frege or, rather, are they all insufficient to clarify the semio-cognitive functioning of thought and mathematical activities?

---

[9] Peirce, C. S. *Collected papers, II. Elements of Logic*. Cambridge: Harvard University Press, 1931.

[10] Frege, G. Sens et référence; Fonction et concept. In: *Ecrits logiques*. Trad. Imbert. Paris: Seuil, 1971.

Peirce	Saussure	Frege
How to classify the variety of representation kinds in the process of interpreting meaning?	What constitutes a language as a common meaning system, despite changes and variations resulting from its multiple uses?	How to explain the strict and non-tautological fecundity of mathematical reasoning?

Fig. 1.6 Three semiotic guiding questions

1.3.1 Saussure: Structural Analysis of Semiotic Systems

The definition of sign proposed by Saussure is revolutionary compared to the classic one. It not only does not include the common property of evoking anything else but also substitutes the notion of the semiotic system for that of the sign. This definition is stated in the following two propositions:

1. The signs have no material reality. They are the invariants of perceptive occurrences that change.
2. The signs are recognized by their relationship of opposition to another one, within a system.

The first proposition is about the *necessary distinction between a sign and its multiple voiced, graphical and semantic occurrences* that change over time and also with the individuals who use the same language. How to recognize the same significant sounds and the same significant units of speech through the verbal productions within the same linguistic community? For example, two different people do not always pronounce the same words in the same way. A sign is not physical; it is its vocal or graphical production by someone that makes it audible or visible.

The second proposition explains the first and is the true contribution of Saussure. The signs can only be recognized as signs through the relationships of meaning opposition they have with other signs *within a system*. Saussure called these relationships, which make up the meaning of the sign "opposition values". In other words, *the sign and its "meaning" are one and the same thing.* Any element can function as a sign only within a semiotic system.

The number systems in basis *n* are the best example to illustrate the two propositions of Saussure (Fig. 1.7).

The two right columns show the occurrence of the same digits "1" and "0", for example. Evidently, they mean numbers solely by their oppositions values to the other digits used in the same basis. Similarly, there is an oppositional value between the successive positions of the same digit. Thus, "10" does not indicate the same number in both columns. These number systems are semiotic systems according to Saussure because their internal oppositional values within a system that is, here, decimal or binary. The two right columns show a world of signs with its intrinsic substitution operations that determine their calculation power. In the left column, on

UNIT MARKS each standing for an integer	Two different SEMIOTIC SYSTEMS using the same characters	
Neither opposition nor position value, but **different spatial arrangements possible**	Basis «10»:**nine opposition values for ten digits** and opposition values for each successive position	Basis «2»:**a single opposition value for two digits** and opposition values for each successive position
I II III . . IIIIIIIII or(III,II,or IIII,I)	1 2 3 . . . 10	1 10 11 . . . 100

Fig. 1.7 Pseudo-objects and semiotic systems

the contrary, the units marks can be traces, stones, coins, with which not any substitution operation is possible, but only external counting or handling of their spatial organization.

It is possible to understand now why the three following distinctions should not be confused: sign and its occurrence, a sign and the object to which it refers to, the signifier and the signified.

- The first distinction is important because it highlights that *it is not the sign that is material, but its occurrence.* This distinction is not only useful in linguistics. We will see that it is also crucial in algebra, for example to understand the development or factoring operations.
- The second distinction is also important because it deals with the difference between the meaning of a sign and its reference to an object. *The meaning of a sign depends on the system in which it functions as a sign.* Instead, the reference to an object depends on an intentional operation designation (Q5 above). This explains the difference between number systems and languages. In number systems the meaning of digits (or sequences of digits) reduces to their designation of integers. In languages, by contrast, the meaning of words depends on their oppositional values to other words, so you must make an explicit and intentional designation operation so that the signs refer to objects.
- The third distinction is secondary since it concerns only the living languages, which are spoken for communication purposes, as shown above. This distinction also refers to the fact that human languages have a double articulation: one phonetic and the other semantic. When Saussure speaks of "signifier"and "signified", he refers to the *articulation of these two types of signs that constitute the oral discourse*: the phonemes and the words, both giving rise to the same structural analysis.

Saussure's model is limited by the fact that its analysis eliminate immediately the diversity of statements that the language allows to produce and the discursive operations that such production requires. Saussure separates the semiotic system that constitutes a language and the use that speakers make of it and is concerned only with the language. When we talk about a "language"in teaching, we are interested in the student's and teacher's discourse and not in the semiotic structure of meaning.

1.3.2 Peirce: The Classification of Kinds of Representations

The Peirce project was to describe the role of representations and signs, in all forms of cognitive activity, from the adaptation to the close environment, to those of scientific exploration. The first question was how to classify the diversity of representations (P1, P2 above), since there is no point in speaking of the role of representations and signs in the different forms of cognitive activity if we do not distinguish the various kinds of representations used. The second question was that, not of their production, but their interpretation since representations as signs are used in a communication process. From this perspective, he resumed the classic definition of sign and added a new element, its interpretation:

"The sign, or representation, is something *that stands* for something *to somebody in some respect or capacity*."[11]

The novelty of Peirce's model is the classification that all types of representation can fill a cognitive function. Its classification is complex because Peirce wanted to do it in terms of a triadic process of interpretation, leading him to distinguish several hierarchical levels of signs. We have retained here only what has been considered the most useful for analyzing a progression in the acquisition of mathematical knowledge by students: the famous trichotomous partition of representations as *a function of THEIR RELATIONSHIP WITH THE OBJECT they evoke* (Fig. 1.8).

This trichotomy is hardly a classification, but a juxtaposition according to two criteria: there is either similarity between *the content of a given representation and the represented object* or causal relationship between the occurrence of a phenomenon and its cause. These two criteria are used simultaneously to distinguish and interpret the representations! They are little discriminatory.

Assuming we can decide about similarity or non-similarity between the representation content and the represented object, what does the use of this first criterion

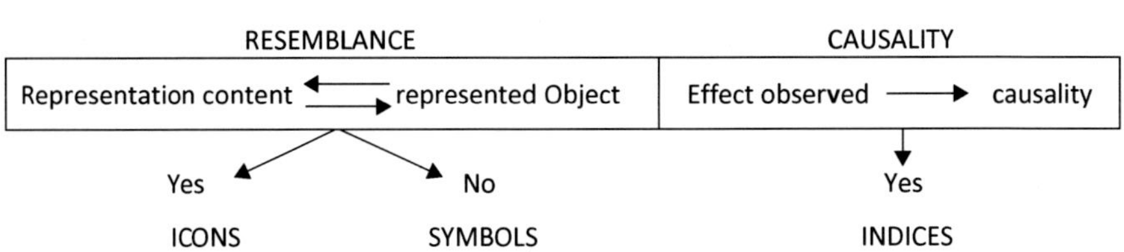

Fig. 1.8 Two or three types of representations?

[11] Peirce, *Collected papers*, 2.228, 4.531.

lead to? To nothing more than the very general opposition between language and image. And, that raises many questions. In mathematics, we must not confuse the statements in natural language, such as definitions or problem statements with symbolic expressions such as equations, formulas or problem statements in the formal language. Should we consider the words of a language, the signs of operations, the letters in an equation, the figures when writing a number as arising from the same kind of '*representamen*'? Similarly, we must not confuse, under the cover of 'icons' or 'pictures', geometric figures, caricatures, sketches, circuit schemes, the geography maps, graphical representations, etc. Moreover, we can ask ourselves whether the photos are iconic representations or indices to the extent that the relationship between the content of a photograph and the photographed object is a causal relationship rather than a referential one. And, in which category all *signals* that command an action like the traffic lights at intersections should be put?

The weak discriminatory character of this trichotomy reflects the definition of signs by Peirce. It is confined to the common property of the representations and signs ("plays the role of ...") and it ignores the specific property of signs (its relationship with the object is a relationship of reference and not cause and effect). In this sense, we can say that Peirce's model still depends on the first knowledge analysis scheme. To confirm it, suffices to remember that his triadic analysis model assumes the cognitive and epistemological priority of the object over all the representations that can be produced. This model rests on a pragmatic approach to what is knowledge and distinguishes three stages in the formation of knowledge, "the primary (*Firstness*), secondary (*Secondness*) and tertiary (*Thirdness*)", that is, the object itself, then a representation of the object, and its interpretation by an interpreter, which finally consists of a new representation of the object and becomes a new object etc.

It is a very strong assumption that removes any distinction between the objects themselves and their different representations. However, mathematics is the knowledge field in which distinction between the objects themselves and their various representations possible is essential. And the teaching of mathematics brutally reminds teachers that the distinction between mathematical objects and their multiple representations constitutes one of the main difficulties of understanding in the learning process.

1.3.3 Frege: Semiotic Substitution and Production of New Knowledge in Mathematics

Unlike Saussure and Peirce, Frege did not propose a definition of signs. And it seems natural because he is in the heart of the semiotic revolution that takes place in mathematics. Signs are the symbols used in Algebra and Analysis for denoting variables, relations, operations, etc. Unlike Saussure and Peirce, he became directly interested *in the semiotic production that might have value simultaneously as proof and new mathematical result*. And to that end, he first developed a conceptual ideography (*Begriffschrift*, 1879).

Frege's contribution comes from his answer to the question formulated by Kant: how to explain that mathematical thinking is a logical reasoning and develops new knowledge (the results of an arithmetic calculation, the theorems of Euclidean geometry).[12] Kant had formulated his question based on the opposition between the analytical judgments necessarily *a priori*, i.e., independent of all experience (tautologies), and the synthetic judgments which, in contrast, are *a posteriori*, i.e., dependent on a given data coming from the sensory experience. And to escape the dilemma he had built, Kant explained that judgements in mathematics were synthetic judgments *a priori*, i.e. based on the *a priori* experience of space and time. Frege takes up this dilemma. But he converts it into the opposition between these two equations:

- *the tautology* a = a
- *the equivalence* a = b, which is the basis of all calculation and reasoning in mathematics (the logically valid deductions).

And, he introduced the distinction between the *signification* of an expression and the *reference* of that expression to explain the possibility of the equivalence a = b. These are the two faces of the meaning of an expression and, unlike the difference between *signifier* and *signified*, THEY CAN BE TOTALLY SEPARATED. Thus, two expressions can have two different significations, but refer to the same object: "3 + 9", "3 × 4", "24/2", "$\sqrt{144}$" etc. *The substitution process that allows calculation is based on the difference between these two faces*: the respective significations or contents of 'a' and 'b' are different, but 'a' and 'b' denote the same object, i.e. in this example the same number. Therefore, we can substitute one to the other *salva veritate,* according to the expression of Leibniz taken up by Frege, to produce new expressions. Here we have the fundamental functioning, explicit or implicit, of mathematical activity and thinking, whether you want to solve problem, to explore in a heuristic way, to prove, or even to apply mathematical knowledge to real situations.

The limit of Frege's contribution is that he considered the symbolic notations and expressions in Algebra and Analysis as the only semiotic representations that meet the mathematical requirements. He did not see the need of other semiotic systems even for thinking and working in mathematics. Thus, he explained that, to make sense and be consistent, the discourse in natural language should function according to the calculation' substitution process, because it is the only one that can be completely checked. This claim goes against the experience of any individual speaking in their mother tongue! And this raised the famous objection by Russell. The proposition "the present king of France is bald", which does not refer to anyone, is simply wrong and not meaningless or absurd.

[12] It should be reminded that Kant totally ignored the development of algebra and analysis, even if he had done the scientific apology of Newton (*Philosophiæ Naturalis Principia Mathematica*, 1686). For Kant, mathematics was the arithmetic and geometry of Euclid.

We can now answer the question that motivates the use of a semiotic approach when analyzing the complexity of the mathematics learning: is there a model that is more appropriate than the others to analyze the student productions as well as the learning activities, in order to acquire mathematical knowledge? Indeed, each of these models considered one not the essential aspects to analyze the role of signs and representations of knowledge in general. But their analyses of representations and signs are not relevant compared to the difficulties to understand mathematics that most students face systematically at all school levels. It is necessary, therefore, to reformulate more accurately the fundamental questions (Fig. 1.6) that led to these models.

The Saussure's question can be reformulated as:

(Q.2) Which discrimination processes allow to RECOGNIZE THE UNITS OF MEANING MATHEMATICALLY RELEVANT in an expression or a semiotic representation?

The Peirce's question can be reformulated as:

(Q.3) Which criteria should be used to classify all kinds of representations that are used explicitly or implicitly in mathematics and mathematics education?

And, Frege's question:

(Q.4) What are the processes of substitution or transformation which are specific to each kind of semiotic representation used in mathematics?

1.4 Conclusion: The Semiotic Representations

We pay very little attention to the central role of representations and signs in knowledge and its development because of their transparency to what they represent or designate. We recognize straightaway the objects semiotically represented and, thus, made present. However, to observe the presence of representations and signs that we believe they provide a direct access to the objects themselves, they should stop being transparent. It is what happens in verbal or written communication for the listener or reader. This also happens systematically in learning mathematics. Whenever we speak of signs and semiotic representations we must remind of their essential phenomenological property that is to reproduce as many times as we want, and according to various systems, some aspect of the same object (above, *P.1* and *P.2*).

Throughout this chapter, we insist on the deep cognitive divide between representations and signs. This divide is almost always little acknowledged and its importance denied. Indeed, in the knowledge activity, the representations and signs fulfill a common function that is "to stand for" what they 'show' or denote and are subject to the same fundamental epistemological requirement of never being confused with the objects themselves. The nature of their respective relationships to the objects themselves is what radically separates the representations and signs. The relationship between representations and the objects results from initial action of objects on physical or sensory receptors. Between signs and objects, there is no causal link. There is just a relationship of reference depending on the semiotic system used, the language, a number system, etc. And that changes everything about the cognitive abilities of treatment, i.e. the transformation of semiotic representations. The transformation of semiotic representation into others is not subjected to any physical constraint.

The semiotic revolution manifested itself with the emergence and fast predominance of equations in algebra, formulas in physics and graphical representations allowing to explore the new curves as the mechanical curves, the folium of Descartes, symbolic notations in Analysis, in short what began to be called "mathematical language", the one that allows us to study nature. For all these productions, it is preferable to speak of semiotic representation instead of "signs." Why?

All semiotic representations have two features that are not found in the elementary units of meaning that we call "signs." First, *they have an internal organization that varies from one kind of semiotic representation to another.* The organization of a simple sentence is not the same as that of an equation. The internal organization of a graphic representation is not that of a geometric figure or a scheme, etc. Secondly, within any semiotic representation, there are always several ways to distinguish the units of meaning and levels of organization. This is obvious for both the sentences and the equations in which the true units of meaning are not the units separated by blanks (words or symbols), but the grouping of units having a meaning different from that of their grouping. This is even more obvious with the geometrical figures, schemes, etc. It is useful to remember that the emergence of the sign notion was focused not on words but sentences, that is, on what is stated. The first designation of the signs was not the Latin term *"signum"* but the Greek word *"λεκτον"* in which we understand the word *"λογος"* that was then translated as *"ratio"*. To summarize, the semiotic representations are, in natural language, the sentences and not the words, in mathematical language the equations and not numbers and letters, in geometrical visualization configuration of figural units and not the points or the straight lines. Too often we associate the signs to the most elementary units of meaning that are just characters to encode units of information: letters, acronyms, numbers, sometimes keywords, or hand gestures. This association amounts to consider the signs as "things" for which we would need first to assign meaning!

What about, then, the cognitive divide between representation and signs for everything that concerns the scientific knowledge and, more particularly, mathematics? This divide becomes fundamental, because it is now specifically the one between semiotic and non-semiotic representations. The first are produced intentionally by mobilizing a semiotic system of representation; the natural language is then the first semiotic system. The second are produced automatically in the mind or by using an instrument. They are the more or less the direct result of object actions on receptor systems.

Annex

Below there is an exercise form to fill. It focuses on the relationships between the fundamental epistemological requirement and the cognitive modes of access to objects.

The exercise consists of filling up this form for mathematics after completing for the other sciences, and finally for the concrete reality, one that is part of our immediate environment. The way we complete this form for these three kinds of knowledge is a test about the manner in which we respond, implicitly or explicitly, the question: *are understanding and learning in mathematics based on the same cognitive processes than understanding and learning in the other areas of knowledge?* (Fig. 1.9).

<table>
<tr><td></td><td></td><td></td><td>Epistemological viewpoint:

Can we distinguish the object from its representation? How?</td></tr>
<tr><td rowspan="3">Cognitive viewpoint:

Which are(is) the mode(s) of access to the objects themselves?</td><td rowspan="2">Physical/sensorial</td><td>Direct perception</td><td></td></tr>
<tr><td>Instruments</td><td></td></tr>
<tr><td>Non-physical</td><td>Representations: which?</td><td></td></tr>
</table>

Fig. 1.9 The separate character of mathematical knowledge

According to the analysis of this chapter, the answers should be:

MATHEMATICS: *no, no, yes (semiotic representations).*
OTHER SCIENCES: *yes, yes, yes (semiotic representations related to mathematics).*
CONCRETE PERCEPTION: *yes, no, might be.*

Chapter 2
Mathematical Activity and the Transformations of Semiotic Representations

The different notions of signs and representations are misleading because they seem to reduce them to simple phenomena of cross-reference to the objects, which signs and representations stand for. The signs and representations would mainly fulfill this function: "evoke what is absent" or "communicate" a thought that is not obvious to others. And, then, we would be dealing with objects rather than with the signs or representations. *However, in mathematics, this function is secondary*. What matters with semiotic representations is first their intrinsic potential for their transformation into other semiotic representations new and equivalent. This is because the power of calculation in development, control of reasoning and the creativity of the mathematical visualization depend on the potential of semiotic systems developed in mathematics and not on the represented objects. In fact, the necessity of semiotic representations in mathematical knowledge covers two very different problems, the non-perceptive access to the mathematical objects and the transformation of semiotic representations into others new but keeping the same denotation.

The first problem is epistemological. Would we have a concrete perception of numbers, functions, etc. or, on the contrary, would the access to them go through, necessarily and immediately, the mobilization of semiotic representations? To support the first hypothesis, the examples of integers and simple 2D geometrical shapes with their topological properties are often put forward. Then, the use of semiotic representations would only become necessary when the magnitude or complexity of these mathematical objects were beyond our limited capacity for intuition or immediate memory, or when we leave the field of the finite and discrete. To support the second hypothesis, there is the fact that this distinction between what would be directly intuitive and what exceeds the limited capacity of our mind is at least fluent and fictitious. In fact, this is based on extremely partial observations.

The second problem is cognitive. It concerns the nature of the mathematical work and, more deeply, the way in which the mathematical thinking functions. Does

R. Duval, *Understanding the Mathematical Way of Thinking – The Registers of Semiotic Representations*, DOI 10.1007/978-3-319-56910-9_2

it consist of the mobilization of "concepts" and the use of common reasoning abilities or, on the contrary, it depends on mobilized semiotic representation systems and the stages of specific thought of mathematics? This issue is cognitive and not mathematical. The difference appears clearly in the analysis of what is called very generally "problem-solving". From a mathematical viewpoint, what we analyze is always the resolution of a given problem and therefore, *we start from its solution* to explicit the mathematical properties that lead to its solution. The analysis is, therefore, retroactive and proper the particular problem that is given or selected, but we can only assume that students will be able to solve all the others problems that are similar. From the cognitive viewpoint, what we analyse is the process that allow us to recognize the mathematical knowledge to be used in the context of the given problem, whatever it may be. For there is no point in explaining the solution to students if each of them cannot see how he/she could have recognized by his/her own the properties to be used. In other words, the cognitive issue is about the intellectual gestures specific to the mathematical work, *even before we have any idea of the solution.*

The question about the nature of mathematical work is not just a cognitive issue. It is also a methodological issue. What kind of observations need to be made and data to be collected for analyzing the cognitive processes of mathematical activity?

- the way students understand or misunderstand the mathematical concepts to be used for solving the given problem? But this remains within the range of mental representations. We cannot read in the mind of the others. We only interpret verbal, graphical or gestural expressions, which are very often more allusive than explicit and their interpretation and cannot be checked.
- the semiotic representations, considered under the only aspect or their "reference to an object"? But, this leads to subordinate them to hypothetical mental representations, which would be the important ones.

We will examine separately the two problems that the mathematical activity raises, the mode of access to mathematical objects and the transformations of semiotic representations. To the epistemological problem, we will present a test about the basic requirement of never confusing the objects with their representations. Is it as clear in mathematics as in the perception of real things? To the cognitive problem, we will see why and how the mathematical way of working must be analyzed in terms of transformation of semiotic representations. *Semiosis* is at the center of the cognitive processes of mathematical thinking through two kinds of transformations of semiotic representation. There is no *noesis without semiosis*, no mathematical thinking without transformation of semiotic representations whatever they are. That is the answer to the true question for everybody who is not a mathematician: "To do maths, what is it"?

2.1 Two Epistemological Situations, One Irreducible to the Other, in the Access to Objects of Knowledge

It is generally claimed that the way of accessing objects of knowledge would be fundamentally the same for all fields of knowledge: first, the experience with the material objects themselves and their iconic representations, followed by the development of their first mental representations and conceptualization. And from this viewpoint, the access to objects in mathematics learning and understanding would be the same than for learning and understanding in botany, chemistry, astronomy, etc. All general cognitive models and some semiotic approaches were built on this assumption.

To check the validity of this assumption we can make a juxtaposition test. It enables to answer two very simple questions.

(Q.1) Can we juxtapose the object itself and its representations?

This seems natural since it amounts to comparing a representation with the object it represents.

(Q.2) When we juxtapose different representations can we recognize whether they are representations of the same object or not?

This means presenting different representations at the same time or in parallel, as we see in encyclopedias, magazines, textbooks, and web pages that sometimes practice this up to kaleidoscopic vertigo.

We can do this test first with material objects and elementary mathematical objects. Would the results be the same to the first question in both cases?

2.1.1 The Juxtaposition Test with a Material Object: The Photo Montage of Kosuth

Look at this photograph taken by Kosuth, in 1965 (Fig. 2.1).

It is the result of two successive pictures. The first picture is that of a chair on which we can sit. This picture is then fixed to the wall, next to the chair and, on the other side, a post explaining the word "chair" is fixed on the wall. The second photo shows this montage. It creates an effect of placing in abyss (*mise en abyme*) the iconic representation of the chair.

The paradox of this photo is that, somehow, it erases any distinction between representation and object, placing the object and its various representations on the same plane, as it is indicated in the caption of this photo: "One and three chairs". This paradox goes against the example considered by Plato to highlight the epistemological requirement on which all knowledge is based on: do not confuse the object and its representation. For trees by a river and their reflections in the water remain separated and quite different as two pages of an open book and, unless we

Fig. 2.1 "One and three chairs": the juxtaposition of the real chair and its representations

are not fully aware, we do not confuse any real thing with its images. However, on the photo of Kosuth the real chair does not overlap completely with its representations because it is in the center, and it is the first element of a setting that falls into the abyss.

The main interest in this photo resides in the fact that it juxtaposes a non-semiotic with a semiotic representation, i.e., a photo and text describing what a "chair" is. We could indeed join other possible semiotic representations in this setting. For example, the mounting template of free chair kits, the arrows to be drawn to connect the different representations between them and the material chair placed against the wall. We may have then "one and *n* chairs" (Fig. 2.2).

First, Kosuth's photo montage illustrates perfectly the two characteristics of the representations (P1) and (P2) we saw in the previous chapter. There are many possible representations of the same object, and the diversity of representations depends on the systems that allow their production.

It is observed that for material objects, there is a direct and immediate access to the objects themselves, and we can juxtapose them with their different possible representations.

The classic cognitive models, as well as the didactic sequence organizations, start from this empirical epistemological situation. They explain *in which order the transition activities between the different types of representation must be organized from direct experience of the objects themselves, in order to make students "construct scientific concepts".*

JUXTAPOSED ELEMENTS	EPISTEMOLOGICAL VALUE of the juxtaposed elements	
1. A **chair** against a wall.	(1) The **object itself** to which we have access independent of its representations.	
	and its different representations	
	NON SEMIOTIC	or SEMIOTIC
2. A **photograph** of this chair.	(2) An **image** produced physically by a photographic camera.	
3. A **text** from the dictionary explaining the word "chair."		(3) A **verbal description**
4. The **drawings are** showing how to assemble the different parts of the chair.		(4) **Another type of image,** *produced by drawing traces.*
5. The **arrows** on the wall are establishing the connections between {1. 2. 3. 4.}.		(5) **A scheme** (*of the "conceptual network" type?*).

Fig. 2.2 The juxtaposition test for a material object

Free arrangements of unit marks into spatial configurations: MATERIALS OR DRAWN ITEMS (PSEUDO-OBJECTS)	Verbal denomination of numbers in mother tongue: ORAL (MENTAL) PRODUCTION OF WORDS	Double internal organization: position and base: invention of "0". WRITTEN FORM OF EXPRESSIONS COMBINING DIGITS
IIII *ou* II II *doigts, allumettes* *la tâche piagetienne* ((((I) I) I) I) *inclusions lineaires* *configuration polygonale*	**"four »** Lexicon varies considerably according mother tongues **"0"** is not pronouced in the verbal denomination of numbers	in the decimal system : **4** in the binary system: **100** in a fraction form of the decimal system: **64/16**

Fig. 2.3 Juxtaposition of "presentations" of an integer

2.1.2 The Juxtaposition Test with the Natural Numbers

Let us now consider a natural number. The question here is whether the first natural are perceptually accessible, unlike larger ones that are only be accessible by using semiotic number systems. Let us taken again the example of integer representation given in the preceding chapter (Chap. 1, Fig. 1.7). We can add the mother tongue words designating integers to the units marks and decimal system. So get the following justaposition (Fig. 2.3).

In this juxtaposition of eight "presentations" of the same number, which one constitutes the perception of the number itself and which presentations would be only representations?

Husserl and Wittgenstein consider that clusters of unit marks are the first perception or intuition of numbers or at least their "proper" representations.[1] Nevertheless, it is difficult to consider the clusters of unit marks as numbers. They only represent numbers for those who can perform *a counting operation*. This operation is complex. It requires some specific conditions:

- *discerning separately and successively* each element of the cluster,
- matching each of all the elements with one word of a sequence of words *always spoken in the same order*,
- attributing to the cluster of marks *the value of the last term enumerated*, which means, according the counting context, either the ordinal or the cardinal.

The unit marks, which can be stones or fingers, *take the place* of any material object. In other words, the counting operation mobilizes THE COORDINATION OF TWO DIFFERENT REPRESENTATIONS: for any cluster of marks units a set of verbal denominations and/or written expressions combining digits. The unit marks fulfill only two functions. They constitute a kind of external memory of the counting, and they provides an immediate synthetic apprehension of the collection of counted material units.

In summary, the first natural numbers are given as objects only in a counting activity of unit marks, which requires an explicit or implicit semiotic production. None of these kinds of representation alone can be considered the natural numbers themselves. The next chapter shows that the same thing happens with the simplest geometric objects. They are not given perceptually but require specific operations, which, to be carried out, go against the intuitive operation of the perceptual Gestalt recognition.

The answer to the first question of the juxtaposition test is problematic. Because the access to the numbers is not direct, but must go through extremely varied semiotic representations, ranging from the most rudimentary verbal descriptions to more complex semiotic systems. This epistemological situation of empirical inaccessibility of mathematical objects is radically different from the epistemological situation of access to all the other objects of scientific knowledge. There is no surprise here, because mathematics begins when we do not limit ourselves to what is given concretely or physically any longer, but when we put it in the framework of what we can conceive as possible.

The second question concerns the recognition of the same object in very different representations (Q.2). Does this recognition depend on the same processes in both epistemological situations? In other words, can we still apply the empirical analysis models of knowledge acquisition to the mathematical objects, whose accessibility is semiotic and not empirical? This question is not theoretical. Any organization of a sequence of activities for student's learning involves, implicitly or explicitly, an answer to that question.

[1]Husserl, E. (1891). *La philosophie de l'arithmétique*. Trad. J. English. Paris: PUF, 1972. Wittgenstein, L. (1983). *Remarques sur les fondements des mathématiques*. Paris: Gallimard, 1983, p. 138–145.

2.1.3 *How to Recognize the Same Object in Different Representations?*

This question translates to the cognitive level the fundamental epistemological requirement of never confusing the representation of the object with the represented object. Two different semiotic representations of the same object can always be taken for representations of two different objects because their respective contents are quite different or, on the contrary, two representations of two differents objects for representations of the same object because their contents are similar. How can we know, then, when we are facing two representations whether they are representations of two separate things or of one and same thing?

The cognitive difficulty results from the fact that two different representations do not have or do not explicitly present the same features of the object they represent. Because even for images, the representation can have been produced as if the object was seen from the front, profile, back, etc. And we know that in the schematics produced for scientific observation purposes, for example, in anatomy or geology, the drawings of the same object can appear to have nothing in common. Instead of images, Frege used as an example the completely different symbolic expressions of the same number.[2] In other words, when we talk about representations, we speak of this complex relationship in which the content of the representation depends on both the kind of the representation used and the represented object used:

{{**CONTENT** of representation, **SYSTEM** of semiotic representation} represented **OBJECT**}.[3]

In the empirical accessibility situation, this cognitive difficulty is easily overcome because the object itself is always accessible outside its representations. Therefore, we can juxtapose it, i.e., associate it with each of the representations that we can produce. Thus, in anatomy and geology, we can always present a specimen of what is represented, i.e., a material object to associate to the different images (Fig. 2.4, left column). There is no more difficulty to recognize it in the different schemes or cuts. The association between the representations and the object itself, the words and the designated things, a work and its model etc., appear

---

[2] We quote two very clear passages from Frege. The first is taken from the article 'Fonction et concept', published in 1891. "The difference of the designations is not a different reason to be different from the designated ... This tendency to not recognize as object what is not perceived by the meaning has as consequence to mistake the signs of the numbers for the numbers themselves, for the true objects of research, in which case 7 and 5 + 2 would be different." The second is taken from the article "Sens et denotation ', published in 1892." We would only know how to distinguish a = a and a = b if the difference of signs corresponded to a difference in the way the designated object is given" (Frege, G. (1971). *Ecrits logiques et philosophiques.* 1 Trad. Imbert. Paris: Seuil, 1971, p. 81, 103).

[3] Duval, R. (2008). Eight problems for a semiotic approach in Mathematics Education. In: Radford, L.; Schubring, G.; Seeger, F. (Eds.). *Semiotics in Mathematics Education; epistemology, history, classroom and culture.* Sense Publishers, p. 39–61.

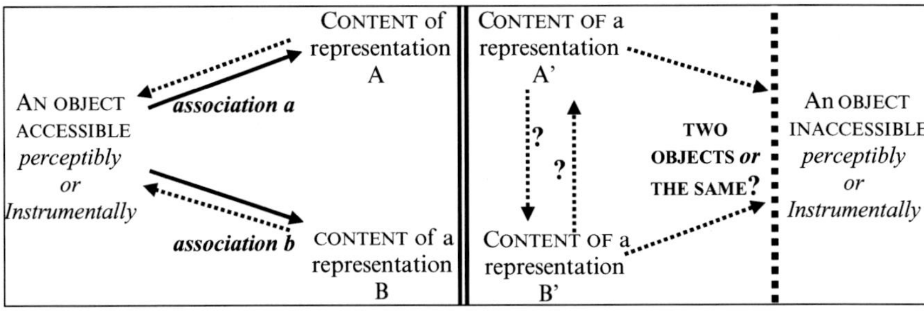

Fig. 2.4 Recognition of the same object in two epistemologically opposite situations

as the fundamental cognitive process to "make sense" and to verify, and hence, acquire new knowledge.

This cognitive operation is not possible in the special epistemological situation of non-empirical access to mathematical objects. We can only juxtapose representations, but never an object and its representation, because the objects of knowledge are not accessible outside the semiotic representations. Also, we need to have a second representation whose content is different from the one of the first, so we can no longer confuse the mathematical object and its representation. But, the question that arises is how can we recognize the same object in two representations whose contents have nothing in common (Fig. 2.4, right column)?

There is, evidently, a local response based, for example, on an operation. To find out whether "3/4" is the same number as "0.75" it is sufficient to divide 3 by 4. But this may be less obvious while finding the corresponding fractional writing to "0.76". However, everything changes when there is no more calculation operation to switch from one representation to another. This is the case when two representations are semiotically heterogeneous, when their respective contents mobilize units of meaning of different kinds (words, symbols, 1D and 2D shapes) and/or when the organization of these units of meanings are of different natures (based either on syntagmatic combinations, or spatial relations and positions). We shall see the importance of these two criteria in the classification of representation registers (Chap. 4). In this situation, the recognition of a single object represented by two representations, A and B is based on a one-to-one mapping between the respective meaning units of the two representations.

Take for example a graduated straight line. It has at least three types of visual units: two kinds of marks corresponding to two scales of magnitude—the first is the division into units, the second is the first division of each unit—and the intervals between the marks for each division scale. We can thus realize an one-to-one mapping between the end of each interval and the sequence of digits 1, 2, 3, … or 0.1, 0.2, 0.3 … We shall see that realizing such an one-to-one mapping between the "number line" and the relative numbers is more complex than is generally assumed. In other words, realizing a one-to-one mapping between the meaning units, which constitute the respective contents of two different representations, is the cognitive condition to be able to recognize the same object in these two representations.

Two questions appear cognitively crucial so that the students understand and acquire the mathematical way of thinking.

(Q.3) How to make one learn to DISCRIMINATE THE UNITS OF MEANING RELEVANT IN THE DIVERSITY OF SEMIOTIC REPRESENTATIONS that are mobilized in mathematics?

(Q.4) How make one become aware of the central role of the one-to-one mapping operation between the meaning units discriminated in two different representations?

Contrary to what has been always postulated in mathematics education, discrimination of the relevant units of meaning in different representations does not result from the acquisition of concepts, but it is the prerequisite for this acquisition. Similarly, the search for the "right" representation or even the juxtaposition of multiple representations are only a misleading help. The "right" representations cannot be associated with the mathematical objects they represent because these are not directly or empirically accessible. The only possible means to access empirically inaccessible objects is to realize the one-to-one mapping cognitive operation between meaning units of the semiotic representations used, whatever they may be.

2.1.4 A Fundamental Cognitive Operation in Mathematics: One-to-One Mapping

We don't have paid sufficient attention to the central role played by the one-to-one mapping elements between meaning units of two different semiotic representations. This operation is the only one that is crucial from both the mathematical and cognitive viewpoints.

The importance of this operation in mathematics appeared with the semiotic revolution, especially with the development of the Analysis of the notion of function. But to illustrate this two-sided operation, we shall consider the simplest and most spectacular example, the historically famous question: are there more natural than even numbers or as many even as many natural numbers? This question seems absurd to most educated people, the inclusion of the even numbers seems so conceptually evident.

The mathematical response consists of one-to-one mapping between natural and even numbers, by paralleling them over two lines to remove the obstacle of the inclusion of even numbers. But this raises the visual obstacle of a twofold occurrence of some objects: any even number occurs in both sequences! It was when they became aware of COUNTING ONLY THE COLUMNS AS MEANING UNITS (indicated by the arrows) *that the students (12–13 years old) had the insight of the*

Paralleling of two sequences of numbers to create new meaning units: the vertical associations of two numbers	*VISUAL ONE -TO-ONE MAPPING between the units of each sequence*	**1** ↕ **2**	2 ↕ **4**	3 ↕ **6**	4 ↕ **8**	.... ↕ ....
oral or mute verbalization	*DISCURSIVE ONE -TO-ONE MAPPING* by counting the new meaning units	↑ One	↑ two	↑ three	↑ four	↑ .....

Fig. 2.5 Twofold one-to-one mapping between two kinds of semiotic representation

one-to-one mapping.[4] We can analyze this discovery as a twofold one-to-one mapping between two kinds of semiotic representation:

- a mathematical one, which leads to the notion of infinity as a set equipotent with one of its own parts,
- a cognitive one, which leaves no trace because it is made orally or even just "mentally" (Fig. 2.5)

Of course, from a mathematical point of view, the first correspondence is the only thing that matters because it breaks the inclusion linked to the representation of the sequence of natural numbers. But, from the cognitive point of view, the second is the crucial one. Without an explicit or implicit counting of the new meaning units shown by the arrows, the mathematical one-to-one mapping between one natural and one even number cannot be understood.

The only cognitive operation for making out new properties, or giving access to mathematical objects is the *one-to-one mapping meaning units from two semiotic representations differing from each other by their respective contents.* The central character of this operation is always ignored because the classical explanation of the formation of concepts *based on the perception of material objects*, and therefore on the differences or similarities of sense data, seems the most obvious acquisition processes for common knowledge. Piaget's analysis of the genesis of the number for the child was the seminal pattern. Although it resorted to this cognitive operation to prepare the famous proof test of the acquisition of number notion, this reduced its importance in two ways. First, the operation is limited to the one-to-one mapping of two rows of pearls, whose spacing of one can vary noticeably (Fig. 2.3, column 1). The invariance of the answers is criterion for the acquisition of the cardinal number notion. And above all, one-to-one mapping has no role in this acquisition, which is explained by two other operations: classification and seriation.

[4]Duval, R. (1983). L'obstacle du dédoublement des objets mathématiques. *Educational Studies in Mathematics*, 14, p. 385–414.

The mathematical and cognitive operations are related to the elements of the respective contents of two semiotic representations. But, the cognitive operation diverges from the mathematical operation in that it cannot objectively defined once and for all, because there are multiple ways to discriminate the meaning units, which make up the content of the semiotic representations. *Its outcome is the recognition of the object represented by two different representations.*

2.2 The Transformation of Semiotic Representations at the Heart of the Mathematical Way of Working

In general, the answer to the question—what does it mean "to do math?" is: "solve problems". Solving problems is, therefore, placed in the forefront of organizing classroom activities. However, this response is in reality vague. It does not say anything about the *mathematical way of working* that should enable anybody to solve problems. Thus, the didactic analysis of solving problems is local and mostly retrospective. It starts from the mathematical solution of a problem posed to explain all the properties that must be discovered and used during the research phase. But, the steps to be performed during this stage are still a black box for many students. To understand the solution when it is explained by the teacher or another student does not allow us to grasp how we should have handled the problem in order to solve it by ourselves. Why be surprised then that many students find themselves back in the same situation of incomprehension or mental block when they face a problem previously explained, but whose the context or one of its conditions has been changed?

The key feature of mathematical of the mathematical work consists of TRANSFORMING THE SEMIOTIC REPRESENTATIONS, given or obtained in the context of a proposed problem, into other semiotic representations. This is where mathematical activity differs from other sciences such as physics, astronomy, biology or geology, etc. This explains why, in mathematics, *a semiotic representation is only interesting insofar as it can transformed into another representation, and not first because the object it represents*. This key feature of the mathematical work involves a complete reversal of the common cognitive viewpoint about the representations and particularly about the semiotic representations. Semiotic representations are not only useful for working with or about the objects. If we want to describe, from a cognitive point of view, the mathematical way of working in mathematics, we must focus on the transformations of semiotic representations and analyze the different kinds of transformation.

2.2.1 Description of an Elementary Mathematical Activity: The Development of Polygonal Unit Marks Configuration

The representations produced only with only unit marks offer no restrictions to any spatial arrangements we want to organize (Fig. 2.3, column 1). Starting with a simple rule, one can with a token, for example, add other tokens to mark the vertices of a square and reiterate … We immediately see that there are two possible mathematical descriptions of polygonal configurations we have produced: one that is the number of elements of each configuration and another, the evolution of the configuration sequences thus generated. Of course, using a literal notation, we can generalize the successive numerical descriptions into a formula.

Here we will limit ourselves to the analysis of the transformations of representations that are involved in a single numerical description activity of polygonal configuration. We will modify slightly an elementary classical situation, by changing the development procedure first, and, then, the shape of the polygonal configuration (Fig. 2.6).

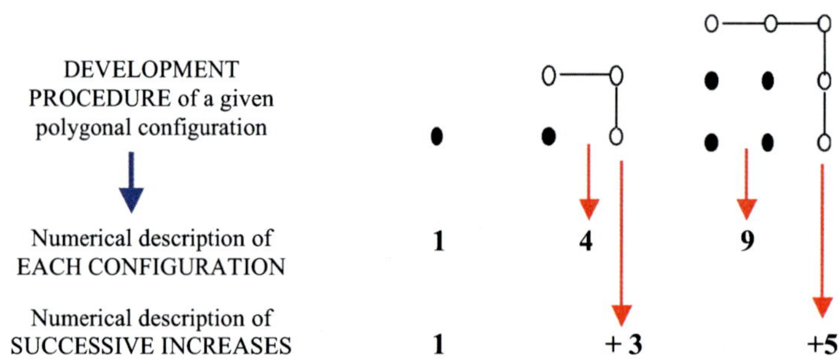

Fig. 2.6 Double numerical description of the development of a polygonal configuration

In the first description, the counted units are the unit marks of each figure. In the second, the counted units are those we need to join in the penultimate figure to obtain the last, i.e., the difference of the unit marks between two successive figures (the white tokens on the figures above). So we have two numerical descriptions possible according to whether one-to-one mapping is made with the square or only two sides of the new square obtained.

We can make a first observation regarding the description of the successive increases. *To carry out this activity with a material, we need to have tokens of two different colors*, one color for the already placed tokens and another for those placed later. If the tokens are of the same color, we must distinguish the tokens added from the previously placed in each successive configurations. The tokens of the same color hide the successive increases rather than show them. We need, therefore, to keep in mind the visual memories of the previous configuration in order to compare it with the new configuration now perceived on the desk, i.e. to shift the focus of

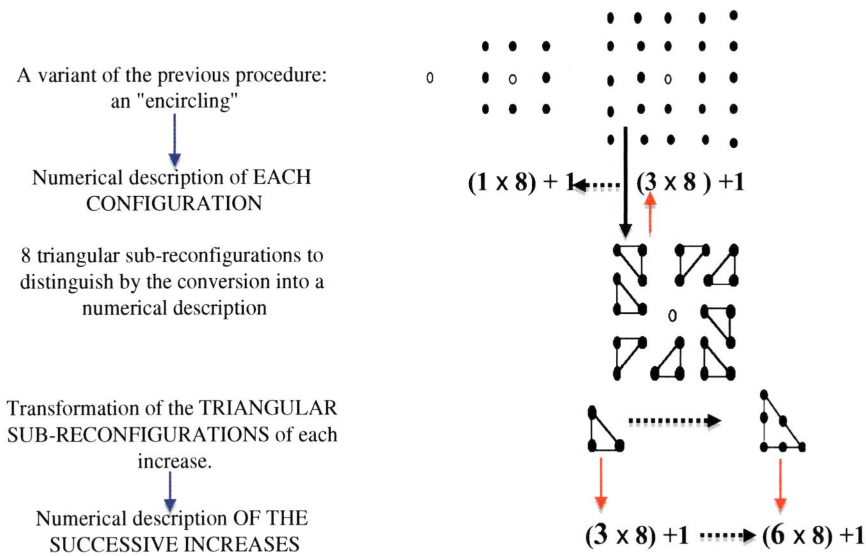

Fig. 2.7 First variant of the development of the polygonal configurations

attention from any configuration to the successive increases! This forces us to wonder about the value of the help that many of the proposed material handlings contribute.

We change now very slightly the transformation procedure. Will the description of the successive increases also be simple? And is it still the same type of problem? (Fig. 2.7).

The counting activity requires to focus on the figural units, i.e., the triangular sub-configurations of tokens instead of the only tokens. But, here, a perceptual conflict arises regarding the recognition of the shape of the figural units to be counted. We need to use as figural units the triangular sub-configurations while the dominant overall shape is a square. We therefore cannot count the two sides of the new square as in Fig. 2.6.

This first variation of the task allows an important observation. Each polygonal configuration is suitable for the distinction of QUITE DIFFERENT FIGURAL UNITS POSSIBLE, but recognizing some of them excludes the correlative recognition of others. The question is, then, to know which will allow to choose from among the many possible figural units, those that are relevant to the task. It is clear that the knowledge of the geometrical properties of the square and triangles cannot help to discern and choose.

Let us apply now this transformation procedure by “successive encircling” by using regular hexagons instead of tokens. As this new task is very complex to carry out without using software, we replace the hexagons by points. We then have a

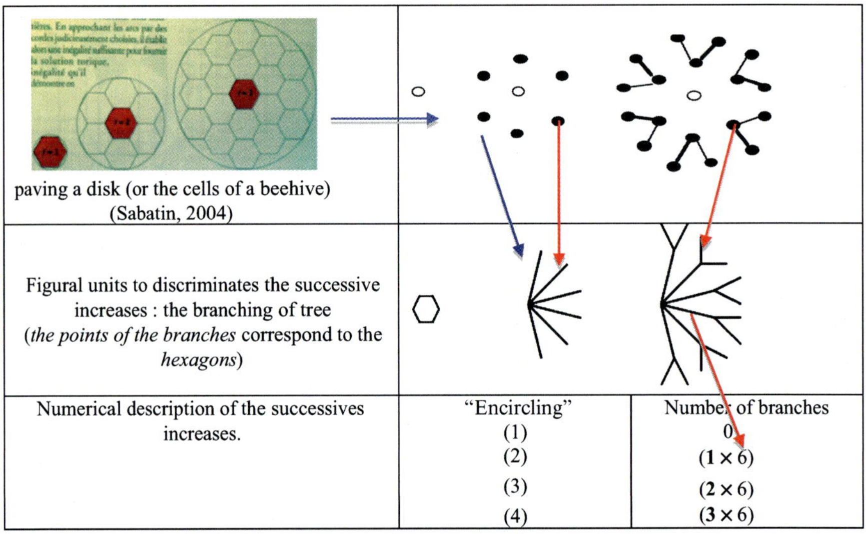

Fig. 2.8 Second variant of the development situation of polygonal configurations

variant of the previous task (Fig. 2.8). From a cognitive point of view, is this still the same task?[5]

Here, the cognitive activity required by this task presents an important difference compared with the preceding variant. First, it is necessary to introduce a transitional auxiliary representation, that of a tree to see how one-to-one mapping between the numerical description and the successive encircling procedure by regular hexagons is possible. It is no longer a "geometric" reconfiguration, because it is not the polygonal configuration and sub-configuration shapes that matter now, but *the points of branching and the number of branches of each point* instead.

The comparison of these three development situations of polygonal configurations shows that a very slight modification of a mathematical task at hand leads to important differences in the cognitive activity required to carry out the task. In the three situations the task is based on the same operation of one-to-mapping (red arrows) between tokens or figural units and numerical expressions. However *two basic kinds of semiotic transformation* are required in the variant situations. The first is the need of an internal shape transformation of the geometrical configuration (Fig. 2.7, dark dotted arrows) before performing the one-to-one mapping (red arrows). The second is the need of introduce quite different representations (Fig. 2.8, blue arrows). Thus, in both variants, two operations of one-to-one mapping, one internal shape transformation (Fig. 2.7) are required to find out the numerical

[5] Sabatin, A. (2004). L'âme de géomètre des abeilles. Les formes de la vie. *Dossier pour la Science*, 44, 72–77

description of the given polygonal configuration development. The initial task (Fig. 2.6.) is more simple because none of these two kinds of semiotic transformation (dark dotted arrows and blue arrows) is required, even if the numerical description asked requires an one-to-one mapping with real objects or marks units (Figs. 2.3 and 2.6).

Whether blue, dark or red (change of the kind of representation, internal transformation of representation, one-to-one mapping between meaning units from two representation contents), all the arrows mark *the cognitive activity that a student must use, either to be able to succeed when solving a mathematical task or to understand the solution.* The first question we ask is whether students facing semiotic representations, whatever they may be, can discriminate the different units of meaning that form the contents of each representation, *in order to recognize the different possible one-to-one mapping with other quite different representations* and whether when looking at figures, diagrams, they can see other spatial organizations besides those imposed by the given configuration. Such transformations of semiotic representations are only required in mathematics and can be truly practiced only in mathematics.

The mathematical way of working can be analyzed here like the transformations, in parallel, of at least two kinds of semiotic representations of numbers, each fulfilling a role different from the other. Some have the heuristic role of exploitation, or intuitive role, in relation to the other, the latter having a description role. But in reality, to make these transformations, *we must implicitly or explicitly go back and forth constantly* between the two kinds of representations. Therefore, they can all perform locally a function of anticipation or control, without the possibility of attributing these cognitive functions, respectively, to configurations of token units or numerical descriptions. At later stages of the mathematical work, it is the numerical description that will be privileged and will allow another kind of transformation: using letters to condense a sequence of local numerical descriptions into a general description.

Is it possible to generalize this analysis based on the variation of a mathematical activities? The two kinds of representations required by the mathematical tasks of polygonal configuration development may seem too narrow. Suppose we no longer practice spatial arrangements about unit marks, but about the signs of decimal numbers. We then have the famous Triangle of Pascal or the Gauss' solution for the sum of natural numbers, with a figural unit count activity that allow a more direct passage to the formulas. And, in the field of combinatorics, we find the same semio-cognitive gestures of the mathematical work. Of course, there is a limitation, we are in the discrete and countable field. As we move to the magnitudes, measurements or the mathematical continuity, do we find the same semio-cognitive gestures? Mathematics mobilizes many other types of semiotic representations. Before studying them, we must revisit the question of the relationship between the semiotic transformations and mathematical activity from another angle.

2.2.2 Representational Transformations Specific to each Kind of Semiotic Representation: The Case of Representation of Numbers

We just saw that the mathematical activity is the transformation of semiotic representations. But, we also found that the mathematical activity can mobilize very different semiotic representations to represent the same objects. We can then ask ourselves if, *in fact, we perform the same operations with the different semiotic representations or, on the contrary, we perform different operations.* The question arises when we switch the kind of semiotic representation (Figs. 2.6, 2.7, and 2.8, blue arrows).

To answer this question, take the example of the representation of numbers, because there is no notion of number without notion of operation that can be carried out on the numbers. Do we make the same calculations when we switch the representation of numbers? This question is only interesting in the elementary situations where the answer is not obvious: the first natural numbers that we can represent either by the unit marks or decimal number system, and the operations with the relative integers.

2.2.2.1 Operations with Small Natural Numbers

There is a fundamental difference between the representations using unit marks and those produced with either the decimal or any other base *n* system.

Representations by unit marks are only a support *for operations that are external to the unit marks:* regroup them or separate them into clusters, arrange them according to the polygonal configuration (Fig. 2.3), or order them at regular intervals over

Two representations of numbers by unit marks	Three operations to carry out addition or subtraction	Symbolic writing of the operations in decimal number system
II III ⇄ IIIII	successive counting of each one of three clusters of unit marks The inversion of the counting order differentiates addition and subtraction	
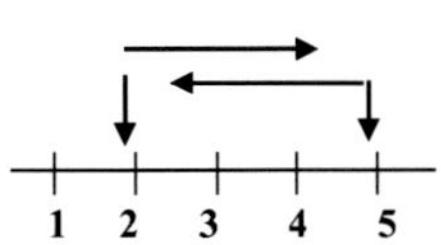	Reading the digit at the starting position, after counting the intervals (or the steps to perform ahead), and finally reading the result at the arriving position	2 + 3 = 5
Two types of significant visual meaning units: POSITIONS and INTERVALS	For the subtractioon: reading the position of the first digit, after counting the number of intervals indicated by the second digit, and reading the result at arriving position	5 - 3 = 2

Fig. 2.9 The additive operations with unit marks

a straight line and encode them, that is, to fix a counting (Fig. 2.9). As either unit marks or digits are used, we do not perform the same external operations to add or subtract two natural numbers from one another.

To add and subtract with representations using unit marks, there are three operations, which are not the same whether the unit marks are associated with digits or not (Fig. 2.9 column 1). And in the case of the number line, the numbers are represented by two kinds of visual meaning units. But in both cases the calculation is not based on any change in the representation of numbers.

However, with the representation of the numbers in the decimal system, *the operations become intrinsic to the representation system* and are carried out by transformations of the digital expressions of numbers. This appears when we consider the numbers that exceed the base system and, therefore, require two digits or more. The limit of the base is marked by this sign that does not refer to a number: “0”. Calculation operations rest on the transformation of two digital expressions that take into account the limits of the base and the position of digits in order to get a third digital expression: 13 + 18 = 31. More generally, a semiotic system of representation of numbers is characterized by its calculation power. The calculation algorithms are related to the operations intrinsic to each system of representations of numbers.

From the representations of numbers by unit marks to the true use of the decimal system, there is the counter-intuitive semiotic threshold of ‘zero’. An always hidden threshold and whose difficulty is underestimated. It resurges when it is asked to multiplying or dividing a number by a factor of 10 or a factor of 100 etc. It is useful to remember that multiplication, division or square root operations cannot be made with the unit marks, but require a representation system involving a double organization (Fig. 2.3).

Therefore, there is a small overlapping zone in which the operations appear to be the same, whether we use a representation of the numbers by the unit marks or a decimal representation of numbers. It concerns the natural numbers that can be represented above the threshold of ‘zero’, i.e., whenever it is not needed to mobilize the calculation possibilities provided by of the decimal system. As if, for example, we would drive a Ferrari only in first gear and just touch the accelerator. For all the numbers in that zone, there is not only a perfect congruence, i.e., complete transparency between the different representations of numbers, but the additive operations may be the same whatever the representation is used. Can we then speak of the intuitive or concrete character of the first numbers and oppose it to the symbolic nature of the knowledge of other numbers, as was done by Leibniz and Husserl, or as it is still done in some didactic approaches? To think so is to forget two essential things. The first concerns the semiotic representation of numbers. The zero threshold varies according to the adopted system. Why, then, a base ten would be more intuitive than a base two, a base seven or a base twelve? The second concerns what cognitively we consider intuition. Simultaneous visual perception of unit marks? But then, can the number vary considerably depending on its disposition or the distance that separates them? The visual perception of a successive series of unit marks? We fall back on the issue of the capacity of short memory, essential for the

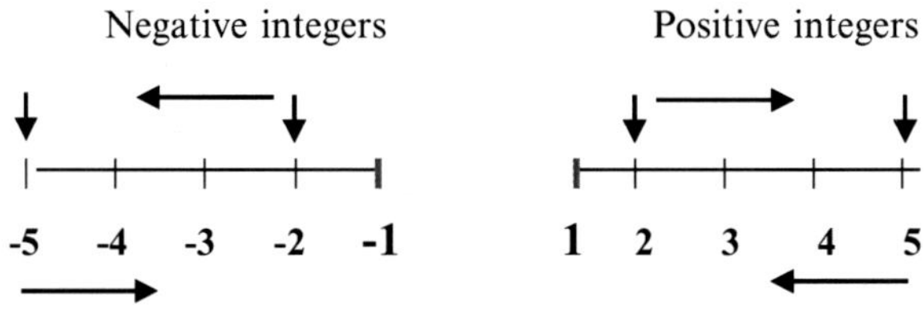

Fig. 2.10 The negative integers as a reverse mirror of positive integers

perception of successive meaning units when listening to a sentence or a phone number for example. There is nothing more uncertain and more variable than the dividing line between what is intuitive and what is not, between what is concrete and what is not.

2.2.2.2 Operations with Relative Integers

The extension of the set of natural numbers to the relative integers requires to encode the decimal notation. It consists of using the operation symbols "−" and "+" for each natural number. This encoding introduces *a new dimension to the value of the opposing sign numbers.* For example, "+1" is not just the opposite of the nine other digits that can be used instead of it, but gets a supplementary value, totally different, that of the opposite to "−1". This opposition is equivalent to reversing the order of unit marks on a straight line, as an image in a mirror (Fig. 2.10).

Of course, this dual representation of the relative integers may seem incomplete. The "0" is missing, which brings together the two representations to constitue what is called the "number line". So the symbols "+" or "−" mean respectively the position value "to the right" or "to the left" of *the origin "0"*. But, would it be that simple? We cannot forget that the number representations are only interesting insofar as they allow calculation. The question is whether the calculation process is the same with each of these two representations, the number line and the encoded digits of the decimal system, or not.

The operations of addition and subtraction of positive integers remain the same as those described above in Fig. 2.9: they corresponds respectively to "steps forward" and "steps backward" (Fig. 2.10, horizontal arrows). It is the reverse for the addition and substraction of negative integers. *But, in both cases we do not get over the barrier of "zero"*. Also, introducing here the "0″ origin to merge the two representations into the "number line" is a cognitive jump that complicates the understanding of these operations.

The numerical line is necessary only when we want to add or subtract a positive integer and a negative integer since it is necessary to overcome the barrier or limit of "zero". But here everything gets complicated because we fall into complete semiotic ambiguity, as we can see it comparing the below calculations (Fig. 2.11).

The semiotic ambiguity is between the position value and the interval value for the numbers, and between the encoding of the decimal numbers and the operations

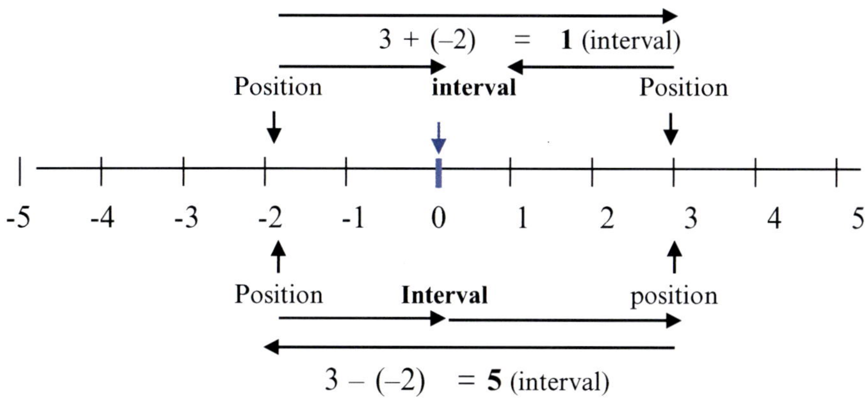

Fig. 2.11 Elementary, my dear student!

steps forwards or backwards for the two symbols "+" and "–" . To add, it is necessary *to start from the position of the positive number* and moving as many steps backwards as indicated by the negative number. The result is the number of the interval between the respective end positions of the two opposite arrows. To subtract, it is necessary *to start from the position of the negative number* and moving as many steps forwards as indicated by this negative number. The result is number of intervals between the positions of the negative number and positive number. It is also necessary to take into account the fact that intervals are oriented. So, if we had considered the operations

$$(-3)+2=\ldots, \quad and \quad (-3) \quad (+2)=\ldots$$

the orientation of the interval is translated as positive or negative number. For addition, it depends on its position on the right or on the left side of the origin "0". For subtraction operation, the value of the interval is always a positive number. Thus, we see the double semiotic ambiguity that the number line arises compared to all other representations.

Let us now look at the symbolic writing of operations. We see, immediately, an increase and diversification *of the possible numerical expressions* for these two operations with the relative integers. Two factors explain them. First, the operation (addition or subtraction) and second, the fact that the operation concerns two integers, both positive or both negative or one positive and one negative. We therefore get eight different numerical expressions possible, that we can represent in a table. To construct such a table it is necessary not to change the absolute value of the numbers and keep them in the eight possible numerical expressions (here |3| and |2|). It is the condition to compare the different possible numerical expressions of the operations and their results (Fig. 2.12).

Two observations are obvious on this table.

First, we obtain the same result in two different ways for each line (A_1 and C_1, etc.) whether performing an addition or subtraction. *The comparison of two numerical*

Operations convertible into steps whether on the positive half line, or on the negative	Operations convertible into steps on the numerical line by overcoming the barrier of "zero"
A_1 3 + 2 = **5** A_2 3 − 2 = **1**	C_1 3 − (−2) = **5** C_2 3 + (−2) = **1**
B_1 (−3) + (−2) = **(-5)** B_2 (−3) − (−2) = **(-1)**	C_3 (−3) − 2 = **(-5)** C_4 (−3) + 2 = **(-1)**

Fig. 2.12 The variations in writing to represent the addition operations with relative integers

expressions on every line of the table shows that subtracting a number is the same as adding its opposite, thus (C_1 and A_1, A_2 and C_2, etc.). It is, of course, *AN IMPORTANT POINT FOR LEARNING ALGEBRA*, since it is this transformation of expression that allows changing the term of an equation from one side to the other. We may also add a third column to the table in which the two numerical equalities of each line are changed into a third. For example, for the first line and the second line:

$$3+2=3-(-2)$$

$$3-2=3+\left(-2\right)$$

So, we have here an example of the transformation mechanism of semiotic representations described by Frege as characteristic of mathematical thinking. *Two expressions whose content meanings are different can be substituted one for another.* On each line of the table, the numerical expressions are obviously different, since they do not use the same symbols of operation and nor does the same relative integers, but they both refer to the same number. We can replace one for the other in any equation in which one of the two expressions occurs.

Second, the two columns correspond to the cognitive jump from the visual representation of the half-lines to the number line. There is no observable difference between the numerical expressions of the columns. But if we look at their respective visual representation support we observe that the visualization of the additive operations with the relative integers becomes very complex. We face a *semiotic overload* of arrows above the numerical line to indicate the position of numbers, the direction of the steps to be done that is always the same as direction indicated par the symbol of the operation to be done, the interval value (Fig. 2.11). This semiotic overload is too often ignored in the didactical use of the number line. Didactic use of number line is so simplified that most arrows needed for indicating all the cognitive operations involved in its support use are missing. Beyond this specific case it raises an important question about the teaching and use of didactic or pedagogical tools: can we use different kinds of representations without blurring the means of visualization or expression transformation specific to each kind, and without short-circuiting the necessary realization of the one-to-one mapping between the meaning units from two representation contents?

2.2.2.3 The Operations with Rational Numbers

There is an important semiocognitive gap between operations performed with the means of transformation of numerical expressions, which decimal number system provides and the ones provided by a graduated line, i.e., by mark units ordered on a straight line from the origin “0”. This gap increases as we move from operations with relative integers to operations with rational numbers.

R. Adjiage showed that, to carry out operations with rational numbers, *one must use TWO GRADUATED STRAIGHT LINES* with *different scales of graduation*: the one-to-one mapping between the position of the unit marks on one straight line and their positions on the other depends on the ratio of intervals of the two graduations.[6] What is essential is that the two graduations will be chosen independently of each other, without keeping the same ratio, as in the recurrent division of an interval. Changing the ratio is here the key point for the organization of learning situations, because fractions are the appropriate numerical expressions for the one-to-one mapping between the position of the unit marks on one straight line and their positions on the other. Thus, for a fixed graduation on the first straight line we may have several other lines each with different ratios and, for the position of an unit mark on this first line, we get different fractions indicating its corresponding position on the other lines 1/2, 3/5 or 5/7 according to the ratio of the graduations. So these different fractions can be compared without calculation, only by relating them to the first graduated straight line taken as a reference. Then we can see *whether a fraction is greater or smaller than the unit interval of the first straight line, i.e., greater or smaller than “1”* and also any other fraction. That does not only define the additive operations, but also the multiplicative ones, which are different from those that can be carried out with the fractional or the decimal writing.

2.3 Conclusion: The Cognitive Analysis of the Mathematical Activity and the Functioning of the Mathematical Thinking

The cognitive analysis of mathematical activity focuses on the problems and the processes of mathematical understanding. But, the comprehension criteria are not exactly the same from the cognitive view point as from the mathematical one. From the mathematical viewpoint, comprehension begins with what is called

[6] Adjiage, R.; Pluvinage, F. (2000). Un registre géométrique unidimensionnel pour l’expression des rationnels. *Recherches en Didactique des Mathématiques*, 20(1), 41–88.

"validation", "justification", "validation", "demonstration", according to the level of requirement. From the cognitive viewpoint, two essential conditions are necessary so that we can speak of comprehension. First, FAST RECOGNITION of the objects themselves through their multiple representations possible, and second, SELF-CONFIDENCE to begin exploring one one's own possible ways in any new task and check their relevance. As long as these two cognitive conditions are not met, whatever is done or explained in mathematics remains for the students a bit like "dark matter". There are therefore two key issues for the cognitive analysis of mathematical activity and the functioning of the mathematical thinking.

The first concerns the access mode to mathematical objects. In this chapter, we took the example of the numbers, but we could have also used the simpler geometric objects such as the elementary figures of Euclidean geometry, or functions etc. This issue is, first of all, epistemological. And, on this point, there is a misunderstanding. When we speak of epistemology, we think *of intra-mathematical* epistemology, essentially focused on the historical stages of the discovery and development of mathematical objects. There are thus epistemological studies of the different kind of numbers, functions, vectors, etc. But, here, it is not what this is about. *The epistemological issue is not intra-mathematical but scientific,* i.e., about the scientific knowledge in the heterogeneous range of its areas and methods. It is whether we have access to mathematical objects, whatever they are, in the same way as phenomena and objects studied in physics, chemistry, geology, astronomy, or biology. In other words, it involves the implicit or explicit choice of a cognitive model of the functioning of thought. Can we use general cognitive models based on an empirical, direct or instrumental, access to the objects of knowledge to analyze the questions regarding mathematics understanding and learning, whereas access to mathematical objects is not empirical, but only semiotic? As the search for examples and counter-examples shows, *the criterion of reality in mathematics is not what is empirically given, but all possible cases that can semiotically represented or constructed.* This issue has nothing theoretical. We have only to observe a student who should successively move from one 50 min session of geography to another of mathematics and then to another of geology or physics, etc. all this within the same day! Not only the objects of study are different, but also and more importantly, the ways of thinking and working.

The second key issue concerns the nature of mathematical activity, whatever the objects, areas or mathematical frameworks. What are we doing when we do math? Obviously to say "we are solving problems" does not answer the question, because this really means "solving *mathematically* problems", which are mathematical problems even if they are presented as real-life problems! In fact, the crucial point is whether an objective description of mathematical activity is possible or not, regardless any claim based on the unverifiable obviousness of introspection. We can formulate it in two ways. What characteristics are observable? Or, what are the intellectual gestures that make you able to work mathematically?

We have seen that the transformation of semiotic representations is the process we find in all forms of mathematical activity. Whether to explore situations, solve problems or demonstrate conjectures, this drives the mathematical activity. The development of mathematical activity depends on two factors: the variety of semiotic representations that can be used, and the need to produce and consider, alternately or in parallel, explicitly or implicitly two quite different representations of the same object.

Taking as an example the most immediately accessible natural numbers, we have highlighted the need to take into account, as a priority, these two factors in order to analyze the cognitive processes underlying mathematical activity.

- the one-to one mapping between meaning units from two semiotic representations differing from each other by their respective contents ist the cognitive prerequisite condition to recognize whether two semiotic representations represent the same object or not.
- We cannot isolate directly the meanings units which make up the content of semiotic representations, and therefore, there are different ways to discriminate them. This depends on the organizational level on which it is focused.
- Some semiotic representations are mixed representations. They result from the superposition or fusion of two kinds of representations, the straight line and the unit marks for a numerical encoding of only some points on the line. We obtain thus many mixed representations: the number line, the graduated straight line (to measure), the real line. In this example, the mixed representations lead to confuse contiguity (adjacency), consecution, visual continuum and mathematical continuum.
- In any semiotic transformation, it is necessary to distinguish the starting representation and new representation produced, i.e. the arrival representation. This raises the question of whether *the inverse transformation is cognitively equivalent to the direct transformation,* that is, whether semiotic transformations are reversible or not.
- There are two kinds of cognitive tasks in mathematical activity. Two semiotic representations differing each other by their respective contents are given or directly juxtaposed. Then you have to RECOGNIZE whether they are two representations of the same object, or not. On the contrary, the semiotic representations that are given do not differ in kind: verbal statements or symbolic expressions, or geometrical figures, etc. Then you have to PRODUCE NEW representations of the same object in another kind of representation. Solving problem always involves, explicitly or implicitly this cognitive task more or less complex.

These key points raise, of course, three questions. What kinds of semiotic representations are used or can be used in mathematics? Do the two factors of semiotic transformations that are at the core of mathematical activity correspond to central processes of the cognitive functioning of thought? What method of analysis (both for the organization of observations and the interpretations of data collected) allows to study the cognitive phenomena related to mathematical understanding and learning? These three questions will be addressed in the following chapters.

Chapter 3
Registers of Semiotic Representations and Analysis of the Cognitive Functioning of Mathematical Thinking

In a semiotic representation, it is not the representation itself that is mathematically essential but all its possible transformations in other semiotic representations. To analyze these transformations, one must take into account the diversity of semiotic representations. The analysis made in the previous chapter shows that they depend as much on the kind of semiotic representation as on its content.

First, the mathematical activity always mobilizes, explicitly or implicitly, two transformations (Chap. 2, Figs. 2.5, 2.7, 2.8, 2.11 and 2.12). The first produces a representation of the same kind as the starting one while the second produces a representation of another kind. We call them, respectively, treatments and conversions. Then, each kind of semiotic representation provides specific means of treatment, which are not equivalent to those of other kinds (Chap. 2, Sect. 2.2). Thus, the procedures to solve a problem change completely according to the kind of semiotic representations used to carry out the mathematical treatment. Finally, the cognitive distance between the contents of the two representations of the same object, but not of the same kind, depends on the nature of the two kinds of representation. Thus, the conversion of one representation into another (to describe the configuration numerically) or the reverse conversion (to produce the configuration that corresponds to the numerical description) may become almost impossible to recognize and to produce (Chap. 2, Figs. 2.5–2.8). The classification of semiotic representations into different kinds and analysis of the semiotic transformations specific to each kind are therefore the crucial issues for the cognitive analysis of mathematical activity, i.e. for the processes of understanding and reasons behind non-understanding in the learning process.

To distinguish and classify the types of semiotic representation used in mathematics is the first step to developing a cognitive analysis tool for the mathematical activities. The problem that arises is whether the representations are to be distinguished according to a direct relationship between the representation content and the represented object, as Peirce did, or on the basis of the production means of representation, which are specific to each semiotic system (Chap. 1, Sects 1.3.1 and 1.3.2). Practically, that is wondering whether the production system of representation would not be a factor at least as important as the represented object in

R. Duval, *Understanding the Mathematical Way of Thinking – The Registers of Semiotic Representations*, DOI 10.1007/978-3-319-56910-9_3

determining the representation content. For example, do a photo and a drawing of the same object have the same content despite some shape likeness? The analyses of the previous chapter clearly impose the second alternative, since, for the same object, the meaning units that make up the content of representations vary according to the systems used to produce the representation.

The concept of a "semiotic system" is not sufficient to account for the two transformations, treatment and conversion, which cognitively drive the mathematical activity. Furthermore, this concept has a double drawback.

It reduces the analysis to an inventory of primitive elements and rules to combine them into new meaning units, rather like formal systems, whereas the transformations of semiotic representations involve complex contents of various meaning units as with problem statements in a natural language, geometric figures, equations, Cartesian graphics, diagrams, tables, etc. These semiotic representations cover two or three levels of organisation of meaning units that merge them into a whole complex content. Transforming these representations, as simple as it seems, require first to discriminate organisation levels and meaning units that are not given separately. And the operations that let you transform a given representation into another do not depend on their content only (as if it were simply by applying the rules) but on the kind of representation. And this discrimination cannot be carried out by applying the formation rules specific to a semiotic system, because they do not allow students not to confuse the object represented with the representation content. There are no trans-semiotic cognitive operations on representations that are produced in one of the various semiotic systems used in maths.

Semiotic systems were assimilated to the codes, that is, to the semiotic systems that have been developed for transmitting information and, more generally, to communicate. This assimilation has reinforced the idea of opposition between semiotic and mental representations, the first serving only to encode the second. And, of course, the thinking processes would depend only on the second. The first consequence of this opposition in mathematics, on which nearly all analyses of the understanding processes are based, is to reduce any change of kind of an object' representation, i.e. its conversion, to a simple coding. This is particularly striking with the Cartesian graphical representations. Plotting them and interpreting are reduced to the rule that associates a pair of numbers to an intersection point on a plan oriented by two graded axes. And that leads to overlook the dramatic drops in students' performances between plotting and interpreting, i.e. between conversion and the inverse conversion! The second consequence is about elementary algebra that is taught in middle school. How to find out the reasons why its introduction and the use of equations for solving problems create an insurmountable barrier to most students, if the semiotic representations are not recognized as independent of mental representations and the only ones to have been taken into account.

The notion of register of semiotic representation has been introduced to account for the two kinds of transformation of semiotic representations that distinguish the mathematical activity from all other forms of intellectual activity. A register is, of course, a semiotic system, but a particular one that runs neither as code nor as a formal system. It is characterized essentially by the specific cognitive operations of which it provides the means to carry out.

To highlight these two points, we will consider the example of language, which is a register and not a code, contrary to the well-known thesis of the structuralism theory. This radical difference appears in the fact that the language consists first of discursive operations of designation, judgment, question, etc., and not of the words of the vocabulary and grammar. The appropriation and practice of language depend on these discursive operations and not the vocabulary or syntax.

To corroborate this analysis, we will consider a second example, the geometric shapes. They are a completely different kind of semiotic representation than that of the language. To determine whether they are a particular register of semiotic representation, we will explicit the figural operations of visualization that the transformation of geometric shapes allows to carry out. These figural operations, which cannot carried out with the other kinds of figures, have a heuristic function in problem solving and give the evidence of invariances which go against any perceptive comparison.

3.1 Semiotic Registers and Cognitive Functioning of Thought

3.1.1 Two Heterogeneous Kinds of Semiotic Systems: The Codes and Registers

We have seen that the signs have no proper sensible reality, be it phonic or graphic. We can only distinguish them by their relations to the all others signs that may be substituted for them with an opposite meaning (Chap. 1, 1.3.1). Therefore, what sets something as a sign or elementary meaning unit is not its local use for the purpose of communication, but its semantical opposition to other elementary meaning units: it is “1” and not “0”, ‘white’ and not ‘blue’ or ‘black’ etc. In other words, any sign is connected to all the other signs of which it is the opposite, thus forming a system that determines their possible semantical uses. It is on the basis of the semantical opposition system to which they belong that signs can refer to objects.

This structural definition of the signs has been a powerful tool for analyzing the diversity of the semiotic phenomena. However, it is not enough to analyze the role of semiotic systems in the development of knowledge because all semiotic systems do not fulfill the same functions. There are those that meet only the *communication functions*, because they allow to transmit the information, or change the physical medium of communication such as, for example, the alphabets that allow to come and go between speaking and writing. These systems are the codes. But there are other semiotic systems that fulfill first, or essentially, the *cognitive functions* of objectivation (becoming aware of something we were unaware), creative thinking

		KIND OF SEMIOTIC REPRESENTATION	TRANSFORMING THE REPRESENTATIONS	CHANGE OF SEMIOTIC SYSTEM
SYSTEMS producers of representations referring to objects *Continuum of meaning*	**REGISTERS** *Languages, figures, graphs etc.*	**A CONTENT ARTICULATING SEVERAL MEANING UNITS** *according to two or three levels of organization*	**SUBSTITUTION BY REFERENTIAL EQUIVALENCE** *Semiotic operations proper to each register*	**CONVERSIONS** by **the one-to-one mapping of meaning units** *non-reversibility*
SYSTEMS for physical transmission of information *Discretization of the information*	**CODES** *Binary code, alphabets etc.*	**SEQUENCE OF CHARACTERS** Each character of the sequence results from the choice of the codification of data (successive states, sounds, ...) and not from the combination rule	Only the **external programming of** the sequences of binary values (Turing machine)	**ENCODING** ↓ ↑ DECODING

Fig. 3.1 Comparison of registers and codes

by using internal transformations of semiotic representations.[1] Thus, numerical systems are registers and not codes as the alphabets (Fig. 3.1).

Unlike codes, the registers of semiotic representation, are cognitively producing systems, or even 'creators' of ever new representations.[2] The production of new representations allows discovering new objects of knowledge. Thus, the graphical representations have created new types of curves, which are not only those obtained with the conic sections but the algebraic curves as the *folium* (1638). Indeed, the content of representations produced by using a register always has two properties. First, it refers or seem to refer to an object. And, it is because of this relationship that the semiotic representations meet a cognitive function. Second, it appears as a complex of elementary meaning units merged into the explicitation of some particular characteristic of the represented object, as seen in the analysis of the previous chapter (Chap. 2, 2.1).

[1] This term appears in *Géométrie*, I, published in appendix in *Discours de la méthode* (1637). It designates the list of signs indicating other signs, so that it can replace the algebraic writing by the geometric figures of conic sections, and vice versa. Two points are essential to understanding the novelty and the purpose of this use:

— Descartes explicitly refers to the semiotic aspect of representations by removing all mental content or any idea that will be associated to it.

— In *Regulae ad directionem ingenii* (1628), he had explained how important it is for the thought to vary and change the kinds of representations (Rule XIV-XVIII). Before Peirce, he had described the diversity in terms of literal written symbols, figures, layouts of unit marks. This work, which presents the model of an algorithmic functioning of thinking, breaks with the conceptual model considered the only possible, to account for the development of knowledge.

The use of the word 'register'crystallizes therefore, the awareness of other cognitive functioning mode of thought, the most powerful mode to mathematicians. This assigns to this word, strangely neglected by historians and interpreters, a theoretical and not just a historical value.

[2] This idea was developed by W. Humboldt, in the early nineteenth century on the subject of language: a language is inseparable from the culture of a people and it constitutes the "body" of thought.

On the contrary, codes are semiotic systems that allow transmitting a discretized information or that commute the encoding of an information according to the means used for the physical transmission (audio/visual, analogic/digital, etc.). They produce long sequences of homogeneous characters, each one corresponding to a "SIGNAL", and these sequences are quickly beyond our capacity of apprehension and short term memory. But, above all they do not refer to anything and, therefore, do not represent anything. The use of code is only a matter of "one character for one information bit" translation of data that may be the state of a circuit, sounds, answers to questions, etc. Coding makes any reference to the encoded data practically useless. The universal code, is, of course, the Boolean code, the one nobody needs to learn and with which the Turing machine was invented, and therefore, computers work. Writing systems are also codes, with the particular ability of merging strokes with either the phonetic production of speech (alphabets) or ideas that any language allows to produce verbally (ideograms*). However, this kind of coding only really works when it becomes automatic or spontaneous.* And here, we find not only the crucial challenge of learning to read, but also its complexity!

However, from the cognitive viewpoint, the difference between register and code does not lie in the greater or lesser complexity of the semiotic systems. It lies in the fact that the registers open possibilities of transformation of the content of the produced representations, which the codes do not allow. As seen previously, in the preceding chapter, it is the transformations of content that produce new knowledge. Thus, the mathematical knowledge does not begin with the semiotic representations of "concepts" or of mathematical objects, but with the transformations of the semiotic representations denoting "mathematical objects". *These transformations depend on the register in which representations are produced.* It is why changing the register is not only to change the content of the representation of an object, but also the semiotic operations to be undertaken to get a new result or prove. The semiotic operations specific to each register used in mathematics are the intellectual gestures needed in every kind of mathematical activity.

We do not analyze in the same way the cognitive functioning of thought when we assume that registers are like codes or, on the contrary, are not codes at all. In the first case, the mobilization of a semiotic system consists of the encoding and decoding of mental contents that would be independent of this coding or that existed before it. Mental contents that are called "mental representations," "concepts," or "information," according to the theoretical frameworks! In the second case, on the contrary, the mobilization of a semiotic system amounts to producing representations referring to the objects and transforming their content in order to develop knowledge. This kind of semiotic activity constitute the thinking process.

Take the example of "language"or, more precisely, of natural languages. Are they codes or registers? This issue is not only important from the semiotic, but also from the didactic point of view. It concerns the role of language in mathematics and its multiple contradictory uses in teaching of mathematics. It also determines how we analyze and use in the educational research, the "interactions" in class or, more broadly, all verbal productions of the students.

The originality and power of natural languages are because they perform at the same time, communication and all the cognitive functions. Depending on whether we favor the communication functions or cognitive functions, we consider the languages as codes or, on the contrary, as registers.

The prevailing answer for a long time was to consider the language a code that allows to transmit information (Jakobson, 1963).[3] It has lead to regarding every verbal formulation as a word encoding of mental contents by the speaker, and its understanding as a mental decoding by the listener. Natural language is not considered otherwise in most research on mathematics teaching. Understanding the problem statements, for example, would be to decode the information that has been encoded into words. And so, 'competencies' of encoding and decoding of the statements have been defined in the evaluation research in mathematics.[4] Similarly, two phases have been distinguished in problem solving, the research phase itself, in which language would not play any role, and the writing phase that would not be more than a matter of formulation, that is, to put into mathematical words. In other words, there would be language on one side and, on the other, the cognitive functioning of thought. The verbal expression would be only necessary to communicate, or fix the result of the thinking work. The problem of this dualistic separation between language and thought is that it allows *neither to analyze the spontaneous and continuous use of natural language by all individuals, nor to describe the specialized use made in mathematics* when defining a property or using theorems, for example.

A language is always used to perform one of these two acts: to say or write something, and to understand what someone else is saying or has written. Usually, we only retain from these two acts the produced speech,[5] i.e., a statement (claim, question, definition, reply, etc.), or a set of statements whose form, status, length, syntactic complexity and logical organization vary considerably. Let us focus for a moment on these two acts. They have absolutely nothing in common with encoding and decoding a message. *To express oneself does not mean to encode an already explicit thought but to objectify it for oneself, become aware, even when we address others.* It is the opposite of coding, insofar as thought emerges only when the words begin to express it. However we have most often the opposite feeling, because, when we speak, we resume or repeat what has already been formulated orally or in writing.

[3] Jakobson, R. (1963). *Essai de linguistique générale*. Paris: Editions de Minuit. See the two studies entitled «Linguistique et théorie de la communication» and «Linguistique et poétique».

[4] The research on national assessment in France has proposed a tripartition of competencies based on the communication scheme: **reception** (read, translate from one language to another), **processing** (apply a technique or a concept), **production** (answer, write a justification). The objectives of acquisition in geometry, calculation, and measurements were defined on the basis of the acquisition of these competencies. Evaluation CE2-6ème Résultats nationaux. *Dossiers Education et Formation*. Ministère de l'Education (1992–2000).

[5] Many researchers in didactics of mathematics speak of "language"and do not care to distinguish between the language and the speech. They use "language"to refer only to the produced speech. This leads to relativize or to jam the differences between the different discursive operations we speak the "mathematical language". (Pimm, D. (1994) Mathematics classroom language: form, function and force In: *Didactics of Mathematics the scientific discipline* Biehler, R.; Scholz RW; Winkelmann B. (Eds) p. 159–169.

Similarly, to understand does not mean to decode a sequence of words or phrases, but to discriminate the meaning units merged at the different levels of organization of the speech and eventually reformulate them. To confirm this, suffice it to recall the systematic and recurring difficulties encountered by students in understanding problem statements. The additive and putting data into equation problems, which we will examine in the following chapters, are the well-known classic examples. There is no rule or procedure to decode or recognize the 'relevant information' in statements. The role of natural language cannot be reduced to a simple communication activity, it fulfills, more or less, the cognitive functions, according to the situations and scientific areas.

Language is not a code but a register of semiotic representation. The wide range of discourses that it allows to produce depends on discursive operations irreducible to grammar or rules. The use of language has nothing to do with that of a formal system. *It consists in performing the discursive operations that fulfill the cognitive functions and that every act of verbal expression mobilizes to varying extents.* What are these discursive operations that each individual should be able to control in either speech or writing situation?

3.1.2 The Three Types of Discursive Operations and the Cognitive Functions of Natural Languages

Two ideas are essential to understanding and analyzing the kind of thinking activity that involves using natural language.

The first is methodological. In all discourse, produced either orally or in writing, we have to discriminate the meaning units merged at the different levels of discourse organization. In order to do this we must analyse the various kinds of speech in terms of discursive operations, because their meaning units are determined by the discursive operation performed to produce each kind of speech. We will see further that the analysis of verbal productions of students should be made based on the discursive operations that are required in mathematics.

The second is cognitive and logic, simultaneously. The fundamental unit of meaning in the case of narration, description, explanation, argumentation, etc., irrespective of speech type, is the sentence. First, the sentence shows the producing power of language. Language allows to produce a potentially infinite number of different sentences, starting from a finite number of words, as Humboldt observed.[6] And any way, we can only understand a language by starting from sentences already produced in this language and those that it will be produced indefinitely. Second, the production of a sentence is not primarily a matter of grammar or syntax: "language refers to a world of objects ... in their complete statements, in the form of sentences"

[6] Coppock, E. (2004). Creativity, generative grammar, and *Erzeugung*: <http: coppock.info/creativity>, Oct. 11, 2004.

(Benveniste, 1966, p. 128).[7] *These observations refer to an essential fact. The utterance of a sentence make us enter another dimension of meaning* than that of words or word combinations to describe anything. The unit of meaning that forms a sentence is constituted by *its epistemic* (absurd, possible, probable ...), *logic* (true, false, undecidable) or *pragmatic* (order, promise, etc.) *value* or *its status* (hypothesis, definition and conclusion) in the progression of a story, explanation, reasoning, argumentation, etc. (Duval, 2007, p. 144).[8] The sentence is the basic meaning unit of speech.

The determination of the discursive operations that underlie the production or understanding of discourse, should, therefore, based on the sentence. We can then determine three kinds of discursive operations: to refer to an object and enunciate something about it, then articulate a new sentence with the one so produced.[9]

The specific discursive operation that determines a sentence is the enunciation of something about something. To develop the analysis of the discourse (*logos*), Aristotle had identified the enunciation with two operations, the affirmative (*kataphasis*) when we enuntiate (*apophansis*) that two things are taken together (*synthesis)* and the negative (*apophasis*) when we enuntiate that they are separated (*diaresis*).[10] But, the enunciation is not limited to two operations, that concern the epistemic value of the sentence. It depends also of the verbs and can, therefore, also be a question, order, promise etc. that concerns the pragmatic values of the sentence.

The enunciation implies another discursive operation: the designation of which we are going to enunciate something. The meaning units corresponding to the description of the objects are not the words but the expressions that combine at least two words. A discourse analysis based on the words (either keywords or mathematical terms) does not provide relevant data, if the objective is to analyze the various ways of describing, reasoning, explaining or more broadly, the acquisition of mathematical knowledge. *The nouns alone say nothing and designate nothing, except for proper names.* Their meaning is due primarily to their opposition to other words that belong to the same lexicon, as shown in the structural analysis developed by Saussure. It is their opposition value in the discourse produced that constitutes the meaning of words. Certainly, a dictionary assigns several different meanings corresponding to the different "uses" of words, but these always refer to the context of sentences that have previously produced in quite different situations. And if mathematical definitions set their theoretical meaning, it does do not remove their value meanings.

The designation of an object always depends on a designation operation. This operation can be rather complex, *due to the insufficient number of words that exist for all objects that we might want to designate.* The designation of figural units,

[7] Benveniste, E. (1966). *Problèmes de linguistique générale* I. Paris: Gallimard.

[8] Duval, R. (2007). Cognitive functioning and the understanding of the mathematical processes of proof. In: Boero, P. (Ed.). *Theorems in schools.* Rotterdam/Tapei: Sense, p. 137–161.

[9] For a detailed analysis of all discursive operations, see Chapter II, «Les fonctions discursives d'une langue», Duval, R. (1995). *Sémiosis et Pensée Humaine.* Berne: Peter Lang.

[10] Aristotle. *De l'interprétation 16 a 12, 16b35-17 a 19. Premiers analytiques* I, 24 a 16–17.

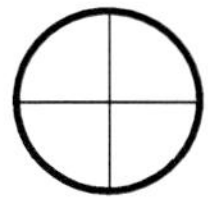

Fig. 3.2 Various verbal designations of one of the five distinguishable 0D figural units

even on the most basic geometric figures, is a school case. So, to designate a 0D unit, makeable in the figure below, we cannot just say "the point" because there are four others visually outstanding. To encode the picture with letters and use the letters is to create a contextual proper name, but it does not solve the problem of verbal designation because no other property can be mobilized without this designation (Fig. 3.2).

In mathematics, we always have even longer verbal designations! Suffice it to read the statements of theorems in textbooks. The key observation for our purpose is that *the discursive designation operations have to fulfill a REQUIREMENT OF UNIQUENESS* in the designation of an object. Also, the didactic challenge of the mastery of the native language is not the knowledge of vocabulary, but the ability to designate many things with the words already known, however the extent of the vocabulary learned. Knowledge of words is nothing if there is no awareness of the designation operations and their complexity. We will find again this question of the discursive designation operation for putting data into equations.

Enunciation opens up the possibilities for expanding to new statements and developing explanation, description, argument or any other kind of speech. Orally, it provokes almost always a question or an answer. *The discursive expansion operations are those that organize a sequence of sentences into a coherent whole according the speaker or writer purpose.* These expansion operations create the cognitive difference between an argument and an explanation, as they create the difference between a mathematical proof and argumentation in a debate or deliberation leading to the adoption of a decision. Thus, it is only at the level of discursive expansion that the sentences have a status, which becomes the most important component of their meaning.[11]

Take the example of a much simpler form of speech, the wording of instructions to construct geometrically a figure. It resembles a description. Consider the figure given to 10–11-year-old students of a school in a tough suburb. They had been previously assigned a similar task requiring the same procedure. To force them to employ mathematical terms, the teacher had written on the blackboard the words that should and should not be used (Fig. 3.3).

Very few students in the class succeeded to designate the figural units. Then, the teacher had prepared after an exchange session on this task, the wording of the geometrical instructions. It consists of describing a simple procedure and its reiteration:

[11] Duval, R. (1995). *Sémiosis et Pensée Humaine.* Berne: Peter Lang. Cap. V: "Le raisonnement".

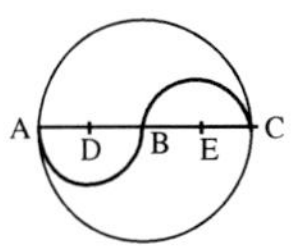

Geometric vocabulary to be used	Words relative to actions that should not be used
Center (of a circle) **midpoint** (of a segment), **half** of a segment, **radius**.	*Compass needle point* instead of center, *Half* instead of midpoint *Pencil lead* instead of the circle

Fig. 3.3 A writing task, to write a message of geometrical construction

1. Draw a segment [AC].
2. B is ***the midpoint of [AC].***
3. Draw a circle ***of center B through A and C.***
4. D is the midpoint of [AB].
5. Draw a circle ***of center D through A and B.***
6. E is the midpoint of BC.....

In each instruction, at least one figural unit must be designated. To designate it, it is not enough to use a geometric term, it is also necessary to situate it in relation to other figural units (here, a segment or a circle). The link between two instructions is provided each time by the double verbal designation of the same figural unit (characters in bold italics). *In other words, it is the discursive expansion mechanism that Frege described to explain the reasoning of the mathematical thinking.*

Therefore, all verbal production has two essential characteristics:

- It includes two levels of discursive organization: the designation operations in the enunciation operation and the successive operations of enunciation in the whole of a speech, whose coherence depends on the discursive expansion operations connecting two enunciations or statements.
- The relevant meaning units that can be distinguished in any verbal production result from the kind of discursive operations. Any verbal production can be broken into discursive segments according to the discursive operations performed.

3.1.3 The Relationship Between Thought and Language: Discursive Operations and Linguistic Expression

These different kinds of discursive operations are cognitive operations. They are irreducible to the application of syntactic rules and knowledge of vocabulary. They are located at the *exact point where knowledge, understanding and awareness*— and, therefore, cognitive condition to develop knowledge—are inseparable. So, to say for oneself and not for the others is the necessary condition to become aware of what otherwise one would not even be aware. The verbal expression paves the way for thought, and not the inverse, as Merleau-Ponty has already observed.[12] Piaget, in his early work, has also observed how important language is to raise awareness of the operations performed.[13] The impression of a separation between thought and language arises from the communication situations in which we express thoughts or ideas that have already been formulated more or less explicitly. Always, without even noticing, we sum up (not to say "repeat"!) in an amplified or reductive manner what has already been formulated, or we have begun to formulate.

The cognitive nature of discursive operations appears in the degrees of freedom that exist between the discursive operations and linguistic formulations produced or used to perform these operations. This fact is manifested in two ways. First, the same discursive operations can lead to use quite different expressions for the same purpose. Second, quite different operations of discursive expansion can use the same linguistic markers or formulations, to perform quite different thinking processes. A typical example is the use of the so-called "logical" connectors in both mathematical demonstrations and arguments, and sometimes explanations. Not only logical connectors do not fulfill the same discursive function in a demonstration and an argument, but the use of connectors required in an argument is useless in a mathematical demonstration even when made in natural language.[14]

Among the three kinds of discursive operations characteristic of a language, the most fundamental is the enunciation of sentences. *Operations related to enunciation are the basic intentional acts of thought, whether this enunciation is oral or written.* The two other kinds of discursive operation are always about the sentences to be produced or previously produced. But, the importance and complexity of these discursive operations is overlooked by the predominance of the oral practice of the language. Because unlike the written production that requires us to perform these three kinds of discursive operations in a controlled manner, the spontaneous practice of speech encourages to short-circuit their realization. Thus, the complexity of the designation operations of objects is always overlooked when we consider only oral production; it is revealed only when a written production is required. It is the same thing for the reasoning modes that depend on the expansion operations specific to each kind of reasoning. Oral production homogenizes into a hypothetical general ability of 'reasoning' the deep and irreducible differences between proving, arguing in a discussion, or even explaining. It is necessary to go through a written production to become aware of the discursive operations appropriate for mathematical reasoning. But, here we are faced with a didactic paradox that we shall examine further. The wording in mathematics requires specific tasks of awareness of the discursive operations proper to mathematical proofs. Otherwise, proving is an impossible task, which turns out to be daunting and useless. Why be surprised then when the introduction of proofs, more or less abandoned, confines

[12] Merleau-ponty, M. (1960). *Signes*. Paris: Gallimard, p. 113–115.

[13] Piaget, J. (1967 (1924)). *Le jugement et le raisonnement chez l'enfant*. Neuchatel: Delachaux et Niestlé, p. 79.

[14] Duval, R. (1991). Structure du raisonnement déductif et apprentissage de la démonstration. *Educational Studies in Mathematics*, n. 22/3, p. 233–261.

itself to *transcribing* a common exchange made orally under the guidance of teachers? Similarly, what benefit to be gained from the wording of instructions to construct geometrically a figure, if specific tasks to realize the discursive operation of designation of figural units have not been done?

The methodological difficulties to analyze the *corpus* of "verbal productions" of the students registered in class or individual interviews result from the ignorance of the basic character of the discursive operations. We tried to analyze the *corpus* according some words used, i.e., surface discourse produced. But this approach leads to a dead end, because it is not relevant to analyze the understanding or non-understanding of the students. It does not provide any controllable criterion to relate the analysis of what the students say with the knowledge of the concepts or mathematical properties. We are then obliged to complement or compensate for what the students do not say or say badly, to interpret their "verbal productions". We think then for the students! Or, on the contrary, we confine ourselves to reproduce long quotations of transcripts, leaving the reader to see how this *corpus* allows justifying the conclusions presented in the article or thesis. The analysis of verbal productions of the students can only be done with a grid built according to the different discursive operations and not based on the presence or absence of linguistic marks or selected words.

3.1.4 Conclusion: What Characterizes a Register of Semiotic Representation

The language is original compared to a code because it is a cognitively creative semiotic system. The creative language power rests entirely on the discursive operations and it is their intentional processes that determine both production of speech and text writing, but at levels of explicitness or condensation that vary extremely from oral to written expression, and with opposite requirements regarding the time cost and control of expression.

All registers are cognitively creative semiotic systems. That means, for a semiotic system to be considered as a register, it is necessary to identify *the specific operations of producing and transforming representations that it makes possible to perform*. These creative operations characterize a register, not the rules of valid combinations for a formal system, or usable signs for a code. Thus, each produced sentence is irreducible to the words she combines. Two sentences can use exactly the same words and not mean the same thing. For example: "The cat eats the mouse" and "The mouse eats the cat" or, if we are disturbed by the epistemic value of the second sentence announcing something unlikely, "The cat causes fear in mouse" and "The mouse causes fear in cat." Refusing the second sentence corresponds to refusing all *cartoons*, all short stories for children and, thus, all products of the imagination! And it is the same situation for the mathematical statements, but without any obvious epistemic value at all! When it comes to deducing a theorem, students do not really see the difference between the theorem and its reciprocal, because their formulations are made with the same words. The only thing that matters is the perceptual obviousness of statements (Duval, 1991, p. 237–238).

Language is the first register of semiotic representation for the functioning of thought. But, generally, language is not considered as such in the teaching of mathematics, in which it is reduced to the communication function. This reduction leads to separate the words and the "information" or "concepts", as if the language were just code . Thus, words are favoured to the detriment of statements, and speech or written expression to the detriment of discursive operations. How surprising is that the way natural language is used in mathematics becomes completely opaque to most students?

3.2 Do the Various Forms of Representation Used in Mathematics Depend on Registers?

Mathematics is the only field where advances in knowledge are closely linked to the invention of new semiotic systems. Their development allowed accessing new mathematical objects: the decimal system and its extensions to access natural, relative and rational numbers; algebraic writing and graphic representations to access functions; the representation in perspective for projective geometry and transformations (for example, symmetry), etc. The predominant use of semiotic systems in all areas of mathematical activity has reduced the use of language to the role of explanation in the margin of mathematical activity or final production of statements. Their differences from natural language seems so important that we can ask whether considering them as registers is relevant. And, above all, what does it bring to the analysis of cognitive functioning of the mathematical activity?

Take the example of the representations produced so that we can "see". They exist in great variety. We limit ourselves to those that seem most natural, the shape figures in geometry. They have three characteristics that give them a particular cognitive power. First, they have an intuitive value as the familiar expression*: "you see on the figure."* They require no further explanation. Then, they enable almost immediate recognition of objects they present as all other drawn images: comic book images (cartoon), sketches, caricatures, layouts, etc. Finally, unlike the other images, they are built instrumentally using either a ruler, a compass or a software because in quick hand drawings, it becomes difficult to distinguish a straight line from a curve and to compare between them lengths or surfaces! This third characteristic requires a way of looking at figures, which is cognitively incompatible with the two first ones.

The use of shape figures in geometry depends on their first two characteristics. The figures allow "seeing" when it is necessary to solve a problem, to demonstrate or apply geometry to reality. Teaching, on the contrary, favors the third feature in order to make the students conscious that the organization of the figures only lies in their geometric properties. Furthermore, the instrumental construction of shape figures, particularly using software, impart a reliability and objectivity to them allowing to use them for heuristic experiments. But here too, for that kind of activity, "seeing" is important because the effective use of a tool requires that we anticipate what can be done and obtained.

Cognitive analysis of the figures concerns *the way they need to be "seen" to be able to use them* to solve a problem or recognize the application of geometric properties in a real situation. Is this the common way of looking at the pictures and see the real objects? Or do we have to subordinate it to a "conceptual knowledge" that it will depend on, and that will lead it? Or, rather, does it depend on the purely visual reorganization operations of the shape figures required for looking at them in a heuristic way? And do these visual reorganization operations depend on the knowledge of geometrical properties represented or, on the contrary, are they independent of them?

All the teaching of geometry is based on the first of these two assumptions and didactic research never questions it, even in the face of the profound learning difficulties that geometry causes. The second is equivalent to assuming that the figures are registers of specific semiotic representation. To show it, it is necessary to describe the purely figural operations that allow, independently of or even before, the use of a geometrical property. These are figural operations that allow to turn any figure into another, to give a solution or produce a counterexample or model a situation. It is the awareness of these figural operations that allows entering the mathematical way of looking at figures in geometry. Here, we present the two most important.

3.2.1 How do We see a Figure?

Seeing involves at least one of the following three operations. The first two are common to the way of looking at the images and the perception of objects. The third is that required in all geometric activity and, above all the only geometrically relevant.

1. Seeing a figure is to recognize at a glance the shapes, i.e., closed contours, which are separated, juxtaposed or overlapping.
2. The recognized shapes may also be simultaneously recognized as being similar to the shapes of real objects. It is this second recognition that is noticed, the first one being transparent like the words spoken or signs we understand. The figure is seen as a more or less schematic image of an object. *It looks like a real object insofar as the neighborhood relations between parts of the recognized shapes keep the neighborhood relations between the parts of the object.* Thus we can define DIFFERENT DEGREES OF ICONICITY. This analysis applies to diagrams, for example, electrical diagrams, blueprints of a building, the maps parts of a city, country, etc. But of course, we cannot distinguish between different

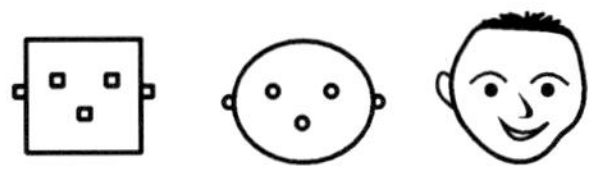

Thus, we can determine the degree of iconicity considering the parts (P) and the global contour (G) of a configuration:

- no resemblance either to (P) or to (Q), but conservation of topological relations between different elements of (P) and those that exist for the represented object.
- no resemblance to (P), but resemblance to (G)
- resemblance to (P) and to (G): we have a representation or a «FIGURATIVE» drawing and not a schema or «ABSTRACT» drawing.

Fig. 3.4 Criteria for the recognition of resemblance between images and objects represented

degrees of iconicity for a photo. Here, it is one of the irreducible differences between semiotic representations and the non-semiotic representations that are produced by a physical system or a device, upon which the represented objects acted (Fig. 3.4).

3. The geometric figures differ from all other visual representations in the fact *that there are always several ways to recognize the shapes or the figural units, even if the fact of recognizing some excludes the possibility of recognizing others.* In other words, to see a picture or a drawing mathematically requires to change the look at the visual representation on the paper or monitor, without changing the given representation .

We must consider the dimensions of the figural units to analyze the cognitive functioning of this change in the way of looking. The figural units that can be recognized are cubes, pyramids, spheres (3D) or polygons, circles (2D) or straight lines, curves (1D) or even points (0D). With respect to points, only the remarkable points (vertices, intersection, and extremities) are visible. The other points should be marked by a coding. In this sense, a figural unit 1D is a visual continuum that represents in an ambiguous way the mathematical continuum of the real line.

However, this is not enough, one must also distinguish the physical substrat of these figural units, such as cube wooden models (3D/3D) or cardboard templates (2D/3D) or the extended wires or a laser beam (1D/3D), from the semiotic representations in perspective (3D/2D) or closed contour surfaces (2D/2D) or plotted curves and straight lines (1D/2D).

Every change of dimension that corresponds to a change in the numerator of the "fraction" (mD/nD) is a considerable cognitive leap, and moving from a physical substrate to a semiotical representation (change of the denominator) is another cognitive leap quite different. I*t is always the figural unit of the higher dimension that is perceptually recognized,* and that blocks the recognition of all figural units of lower dimension, because it merges visually all these figural units potentially involved. *Seeing "geometrically" a figure is to operate a DIMENSIONAL DECONSTRUCTION OF THE SHAPES* that we recognize immediately into other shapes that are not seen at first glance, and this without changing anything in the figure displayed on the monitor or paper. In the examples below, we limit ourselves to the plane geometry figures taught in elementary and middle school (Fig. 3.5).

Figures 2D/2D **Global configuration**	Two decompositions VISUALLY INCOMPATIBLE into **figural units 2D**		And a third into **figural units 1D**
	Fitting together by JUXTAPOSITION	Fitting together by OVERLAPPING	DIMENSIONAL DECONSTRUCTION OF SHAPES
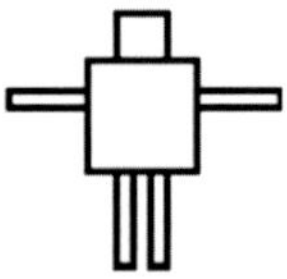	**6** SHAPES (*closed contours*), each one being a part of the human body ICONIC	**5** SHAPES, narrower or wider rectangular bands NON-ICONIC	**18** SEGMENTS or **12 SUBJACENT STRAIGHT LINES**
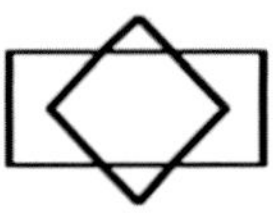	**5** POLYGONAL SHAPES (*2 triangles, 2 pentagons, 1hexagon*) NON-ICONIC	**2** REGULAR POLYGONAL SHAPES (1 *square* and 1*rectangle*)	**8** EDGES OF **UNITS 2D** (*sides*)

Fig. 3.5 Three ways of recognizing figural units *n*D in a plane geometric configuration

We can easily observe that recognizing the 2D shape figural units by juxtaposition and overlapping in the first and second pictures, respectively, *is immediately imposed to the sight.* This prevents us from recognizing the other 2D figural units possible, and particularly the 1D figural units. However *the mathematical way of seeing the figures in geometry requires the possibility of moving from one to the other spontaneously and quickly*, because all definitions of geometrical properties require the recognition of the figural units 1D and 0D, i.e. the points that are the intersection of two figural units 1D (here the vertices or line intersections).

There is, therefore, a considerable cognitive leap between the normal and the mathematical way of seeing. In the usual way of seeing, we do not ever take into account the dimension of the figural units we recognize and we are not concerned with varying this dimension to recognize other figural units we do not see, but which will become more important than those we see. *This is a variation which needs to be done in the look at the figure, because otherwise it never possible to recognize that must been seen according the geometrical statements. Naturally, this requires a long training because it goes against the automatic operation of the perceptual recognition of shapes.*[15] Which student, used to read comic books or to orient himself/herself on a plan, can him or herself guess this radical change that requires the use of figures? It is not in the technology or geography classes that we learn to look and use the geometric figures. The statement of properties in the definitions and theorems are based on this violent and unrealistic way of seeing that.

[15] Duval, R.; Godin, M. (2005). Les changements de regard nécessaires sur les figures. *Grand N*, n. 76, 7–27.

3.2.2 The Two Types of Figural Operations Proper to the Geometrical Figures

A representation is only interesting in mathematics as it can be transformed into another representation (Chap. 2). This, evidently, applies to the figures. They give rise to two kinds of figural operations. There are those that are based directly on perception to transform 2D/2D figural units (or 3D/3D objects) *into others of the same dimension*. They are those developed since the beginnings of geometry. They have the particularity of being realizable by working with material objects 2D/3D or 3D/3D. And, there are those of the dimensional deconstruction. No manual handling of objects nD/3D can simulate them. They appeared later in geometry history.

1. The mereological division of a global configuration *into figural units of the same dimension* (2D→2D) and their reconfiguration into another whose overall contour is the same or not. This operation constitutes one of the major heuristic transformation of geometric figures. It was used as a convincing proof of mathematical properties (Fig. 3.6).
2. *The DIMENSIONAL DECONSTRUCTION of shapes* ***(n D → (n−1) D)*** *allows to analyze the transformation of a given shape into another shape of the same dimension even if it seems completely different.* The enunciation of the properties justifying this transformation *considers the figural units of the next lower level:* secant planes for 3D figures, networks of straight lines for the plane figures, or even pairs of points (a point and its image) for segments, curves, etc (Fig. 3.7).

Therefore, the first main activity for learning geometry is not to construct figures instrumentally or with a software, but to deconstruct dimensionally all the recognized shapes 2D. This requires a specific operation, which has become reflex for mathematicians and teachers, but it is not all for students at all: to systematically extend all the segments so as to leave the closed contour encompassing the figure, and extend them until reaching their intersection points that will enable us to draw new lines. The classic problem shown in Fig. 3.8 provides a perfect illustration.

Then it will be remain to recognize the inscribed quadrilateral of the Figure A in the network of lines drawn and erase all the superfluous lines. It is the same geometrical transformation of a configuration into another as in the previous example wherein the stage 3 is the representational mainspring of all figures not involving any curves (Fig. 3.7). Recognizing quite different shapes 2D in any network of lines is the crucial cognitive operation for becoming aware of the dimensional deconstruction of shapes 2D. And this dimensional deconstruction is involved in every definition, in all reasoning as in any explanation regarding geometrical figures.[16]

---

[16] Duval, R. (2005). Les conditions cognitives de l'apprentissage de la géométrie: développement de la visualisation, différenciation des raisonnements et coordination de leurs fonctionnements. *Annales de Didactique et de Sciences Cognitives*, n. 10, 5–53.

Duval, R. (2004). Como hacer que los alumnos entren en las representationes geométricas. Cuatro entradas y... una cinqua. In: Lopez Ruiz, Jesus (Ed.). *Números, formas y volúmenes en el entorno del niño*. Madrid: Ministerio Educación y Ciencia, Secretaría General Técnica, p. 159–187.

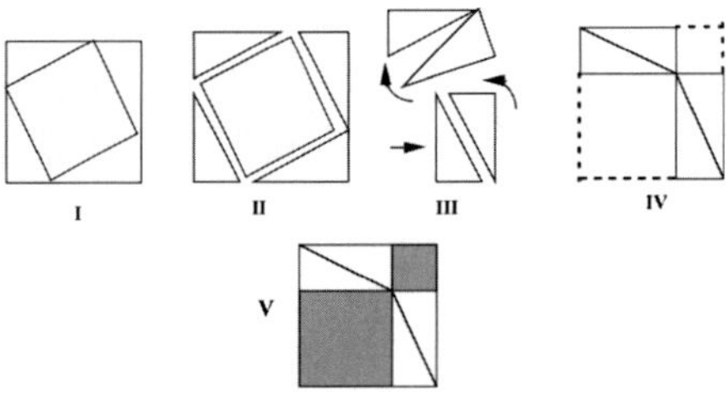

The mereological division of a 2D convex configuration into several 2D figural units 2D (I and II).

The reconfiguration of some figural units (III) into another concave configuration (IV) while making the reconfigured units invariable.

Overlapping of the convex contour including (IV) with the starting configuration (I) to show another equivalent configuration (V)

Fig. 3.6 Mereological transformation of a configuration of figural units 2D into another

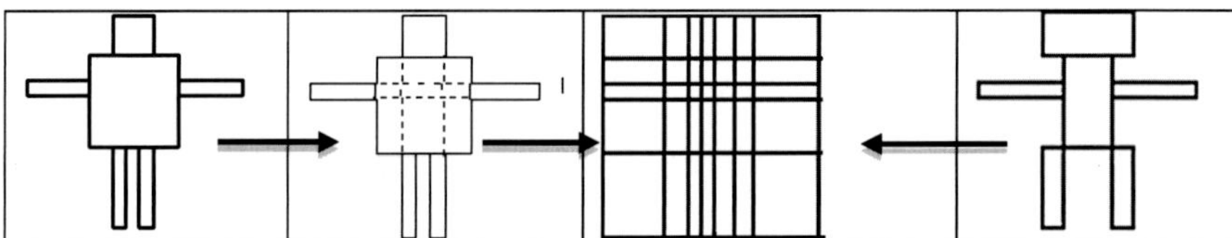

1. A fitting together of 2D shapes that look like a man.	2. Internal extension of figural units 1D	3. Deconstruction of the shapes into the subjacent network of straight lines (figural units 1D) common to all possible recognition of 2D shapes	4. Recognition of another fitting together of possible of 2D shapes that look like a man

Fig. 3.7 Figural operations allowing to transform geometrically a figure into another

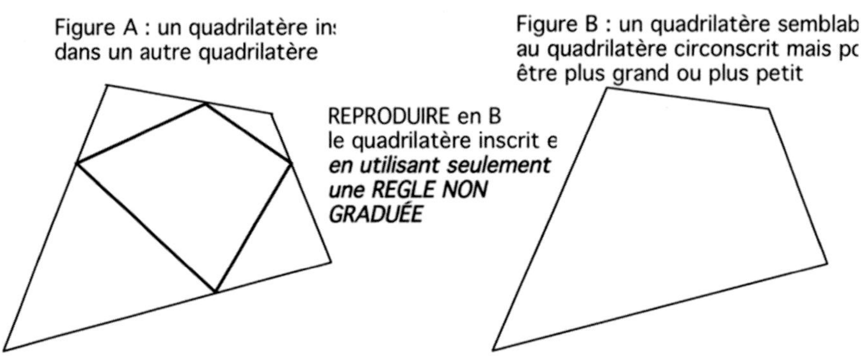

Fig. 3.8 A problem of construction requiring a dimensional deconstruction

3.2.3 Concealment of the Register of Figures in the Teaching of Geometry and Didactic Analyses

All heuristic use of figures in problem-solving, every explanation of a geometric property with the help of figures or even, for some, with the handling of a material, all wording of the statement of properties with a figure to justify or prove a conjecture depend only and exclusively on these two figural operations of Section 2.2. Without real awareness of these two operations, students and educated adults can only remain blind to the geometric representations and expect that every time someone will tell them what to do. And yet, the teaching of geometry remains organized in unawareness, or censorship, of these two operations that make the figures, a COGNITIVELY PRODUCTIVE SEMIOTIC SYSTEM OF VISUALIZATION. Why this situation?

The opposition often made between "drawing" and "figure" is symptomatic of this situation. The "drawing" is the particular configuration shown on the paper, on the blackboard or on the computer monitor, while the "figure" would be the object properties represented by the drawing or still, the class of all drawings that may be the visual representations of the object. This opposition has taken on a very strong meaning with the "dynamic geometry", illustrated by the following non-animated mini-sequence (Fig. 3.9).

We can wonder about the cognitive relevance of this distinction. It involves two serious consequences for the analysis of learning processes in geometry and the difficulties encountered by students.

It suppresses the important difference between the manual drawing (Fig. 3.4) and *the instrumental drawings, each of which produces a specific regular line according to a geometrical property* (a "straight" line, arc, circle). These specific regular lines distinguish the geometrical figures from images or diagrams.

It suppresses the importance of looking and visualization. *The figure is identified with the properties we do not see because no drawing show them in their generality.* These properties can only be learned by concepts, that is, terms defined in statements. This reflects, in fact, the conglomerate of instrumental and semiotic productions called "figure" in teaching. It consists in subordinating what is given to see by the drawn configuration to verbal indications fixing as hypotheses the geometrical property or object to be recognized. For we can never be assured whether what seems perceptively obvious is the geometrical property "seen" (parallelism, perpendicularity, midpoint, symmetry, etc.), or not. And for the same drawn configuration,

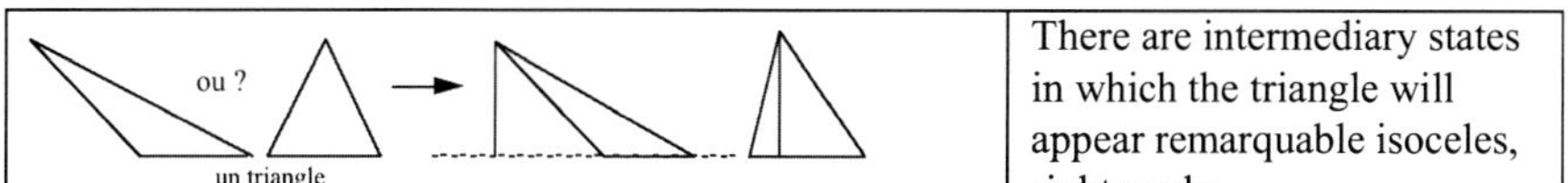

Fig. 3.9 A "figure" as a variation of "drawings"?

we may have different hypotheses. However, we assume that the same drawn configuration, will be seen differently by the students! In other words, we assume that students will look at the configuration produced with the verbal glasses of the given hypotheses. Such a semiotic conglomerate can be a simple phenomenological object under the eye of an expert. But, it is not what happens to students even after several years of learning.

It is necessary to be aware of the figural operations and to have acquired the reflex of the dimensional deconstruction to recognize the multiple figural units merging in the immediate recognition of any 2D shape. The opposition between drawing and figure ignores the complexity of the mathematical way of seeing, which is cognitively required to use a mathematical reasoning for what is given to see. This might be explained by the fact that the figural operations are not associated with any particular concept and they escape all decomposition into particular knowledge prerequisites, while all educational programs are mainly based on the acquisition of concepts and competencies for using each concept. Which are the learning conditions to promote the geometrical visualization that is the first threshold to cross for understanding geometry?

First, we must propose tasks from which all measurement and calculation activities are excluded. To learn how to see, students must learn to work without resorting first to metric aspects. The internalization of figural operations is always the condition necessary to be able to do numbering tasks, apply either the area and perimeter calculation formulas or a geometric property in various real situations, such as the Thales' Theorem. Thus, we have seen that the numbering activities related to the regular growth of the elements of a configuration require carrying out figural operations: *dividing a square configuration into triangular sub-configurations* (Chap. 2, Fig. 2.8)!

Secondly, the organization of tasks cannot be the same for mereological reconfiguration and dimensional deconstruction operations.

The mereological reconfiguration operations are based on perception and play with the recognition of the 2D shapes. The simple perceptive recognition of the figures can be of help or, in the contrary, an obstacle for solving a problem. It depends on the chosen hypothesis and verbal formulation of the question asked. It is necessary to organize the tasks *by varying the figure, from the situation that helps to "see" the solution to these where it becomes difficult or impossible to see it.* We can thus identify figural factors that facilitate figural treatment of the problem and those that, on the contrary, inhibit. And we were able to measure their impact on solving very different problems. These figural factors constitute a tool to evaluate *a priori* the cognitive cost of the various figural proofs such as those of the Pythagoras theorem and whose the classic presentation (Fig. 3.6) shows the simplest case.[17] These factors are also important didactic variables to organize a heuristic learning for solving classic or basic problems in geometry.

Organizing tasks to become aware of dimensional deconstruction is much more complex. Dimensional deconstruction of shapes goes against perception, that is,

[17] Duval, R. (1995). Geometrical pictures: kinds of representation and specific processing. In: *Exploiting mental imagery with computers in Mathematics Education*. Sutherland, R.; Mason, J. (Eds.). Berlin: Springer, p. 142–157.

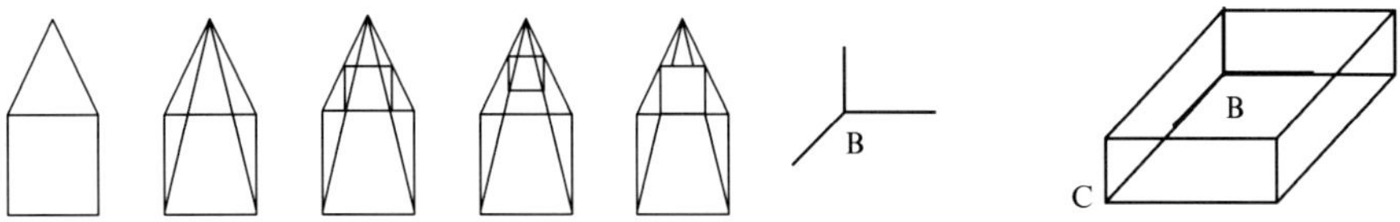

Fig. 3.10 Mobile or fixed focus of the dimensional sight

against the immediate recognition of 2D/2D or 3D/2D figural units that imposed themselves at first sight and block the recognition of other figural units. This occurs even when the lines seem separated, because it is always the figural unit of the higher dimension that is obvious at first glance and merges the other units into itself, as the Gestalt theory has shown (Fig. 3.10). The solution of a geometry problem 'in space' requires necessarily a dimensional deconstruction operation, i.e. *seeing the 2D form obtained by the intersection of a solid with any plane in space,* and not some spatial ability to see 'in space'. This requires specific a specific range of tasks extending from handling both 3D/3D and 2D/3D materials to their multiple representations 3D/2D and 2D/2D.[18]

The cognitive problem is different in plane geometry. Visual perception systematically imposes the recognition of 2D figural units against all recognition of 1D figural units, regardless of whether they belong to the perceptively recognized 2D figural units or not. *It is this dimensional deconstruction that is required by the mathematical discourse in the definitions of polygons.* Practically that required a reflex explicitation of the network of subjacent lines to any 2D figure given and the recognition of all figures 2D possible with their diagonals. To make the student enter this way of seeing it is necessary to develop specific tasks and problems starting at primary school level.[19] Without it, we create the conditions for a cognitive schizophrenia between the elementary figures introduced as dried plants in a herbarium and a vocabulary that requires, on the contrary, the dimensional deconstruction of 2D shapes, and will not be used by the student to formulate anything meaningful to him/her and mathematically acceptable.

3.2.4 Geometric Visualization and Problems from Reality: Direct Passage or Need for Intermediary Representations?

A major change happened in geometry teaching during the last fifteen years regarding teaching objectives and the way of introducing geometrical knowledge. The aim is no longer to understand the demonstration, *but an empirical approach of*

[18] Rommevaux, M.-P. (1998). Le discernement des plans dans une situation tridimensionnelle. *Annales de Didactique et de Sciences Cognitives*, n. 6, p. 27–65.

[19] Duval, R.; Godin, M. (2005). Les changements de regard nécessaires sur les figures. *Grand N*, n. 76, p. 7–27.

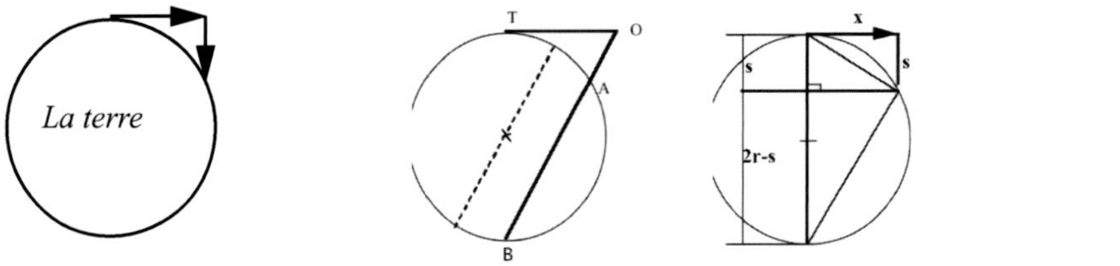

1. SCHEMATIC IMAGE:
the circle: *the earth*
the horizontal arrow: *a projectile launched at a velocity x per second*
the down arrow: *its fall s towards the earth*

2. GEOMETRIC VISUALIZATION OF THE THEOREM:
the product OA.OB is constant,
or OT = √ OA.OB

3. SUPERIMPOSITION OF SCHEMATIC IMAGE ON GEOMETRIC FIGURE
leading to the formula:
x = 2rs
r : *radius of the earth*
s : *distance of the fall per second*

Fig. 3.11 Semiotic representations, imagination and scientific creativity

properties and concepts in the context of problems corresponding to real situations. It is expected, therefore, not only to motivate the teaching of geometry but make the geometrical objects more accessible to the students. And it is assumed that the figures will more easily be seen as a representation of relations that we observe in reality and a modeling for the concrete problems encountered.

The application of geometric knowledge in physical reality raises a cognitive and epistemological problem that is crucial for the teaching of mathematics. It appears when we start from a real problem that has not yet been modeled geometrically. Do we go directly from the data recorded in a real situation to the geometric visualization or do we need an intermediary representation? To analyze the cognitive processes of this passage, we will consider two examples.

We know that to explain the orbit of a planet in relation to the Earth, Newton used the power of a point theorem.[20] In this explanation, the theorem does not apply directly to the described physical phenomena, but to an image that schematizes them. This image plays an essential cognitive function: it allows *a transfer by analogy* that enables to see the Earth/Moon system as the relationship of a circle in relation to a point. For this, a semiotic condition is required: the *schematic image and the geometric figure corresponding to the theorem must be congruent* so that the two representations may be superimposed and appear to be the same. In other words, the jump from physical reality to geometric model is not direct but requires another kind of representation used a bridge between them (Fig. 3.11).

The geometric modeling of a concrete situation requires A MIXED REPRESENTATION THAT RESULTS FROM THE SUPERIMPOSITION OF TWO DIFFERENT SEMIOTIC REPRESENTATIONS:

- A sketched picture of significant data of the physical situation that allowed a reasoning by analogy.

[20] Feynman, R. (1979). *Mécanique 1*. Trad. G. Delacote. Paris: Inter Editions, p. 89–90. Feymann gives only the third visual representation of the Fig. 3.11, and explains the reasoning by analogy leading to it.

- A figure whose figural units may be placed in one-to-one mapping with the figural units of the schematic image, and which will show the geometrical relation to be used.

These two superimposed representations (1 and 2 on the figure above) have quite different cognitive functions. Thus, *the schematic image can be converted into a verbal description of phenomena that have nothing in common with the mathematical statement of the theorem properties.* This leads to an important consequence for teaching. Giving immediately the mixed representations or limiting only to geometric figures, even in an empirical or pragmatic approach, increases the difficulty of the cognitive gap between geometric 'knowledge' and the real situations in which the students will be called to apply them. *Activities about the production of schematic images and their articulation with the figure of a theorem are as fundamental as those about calculations that must be performed to apply a theorem.* And what is at stake here from an educational point of view is the development of the imagination and innovation abilities for applying geometrical knowledge in any situation.

The second example is essential for teaching. It is the use of Thales' Theorem. Currently, many textbooks present mixed representations, which superimpose a coded geometric figure on a schematic image. The difficulty is that there is not only one real situation of application, but three, *and therefore, three totally different schematic images.* Similarly, there is not one possible geometric shape that can be linked with the statement of the theorem, but at least two, as we shall see below (Fig. 3.12).

Therefore, there is not one typical mixed representation for applying this theorem, but six. And, making the correspondence between each of the schematic images and the two possible configurations of similar triangles works for some superimpositions but not for all. We see, therefore, the cognitive complexity of an empirical approach or a pragmatic use of this theorem. It does not lies in the difficulty of calculating ratios, but in the articulation of these two kinds of figures, i.e. in the recognition of the ratios relevant in a given situation. It is their cognitive articulation that allows to see which distances should be measured in a real situation to calculate the inaccessible one that cannot be measured.

But almost always, in education, we propose to work straightaway with the mixed representations. However they have nothing concrete or empirical because we do not care to make them vary according to the real situations possible. We leave to the students the responsibility to guess the different situations and the two kinds of representations to be articulated, without which it impossible to use the theorem outside a classroom session.

3.3 Conclusions

The mathematical activity consists in transforming semiotic representations into other semiotic representations to get new information or knowledge and solve problems. The notion of register is necessary to analyze the cognitive functioning

	ICONIC VISUALIZATION	GEOMETRIC VISUALIZATION
IMAGE, PLAN, PHOTO...	SCHEMA of a situation of visualization of a point *to perform the accessible measurements...*	Geometric configurations
Shows the RESPECTIVE POSITIONS OF THE REFERENCE POINTS over a place or scenery (properties, lighthouses, distant objects, boat) *in order to recognize the three possible meausures to be done*	Aiming at a top point from the ground: 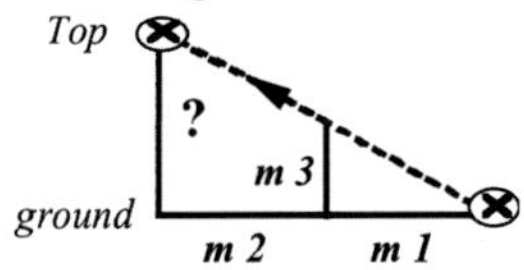 Aiming to a ground point from the top : 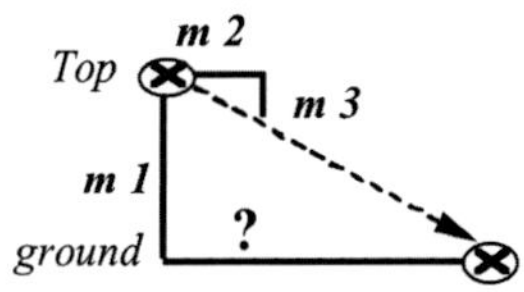 Aiming from two distant points : 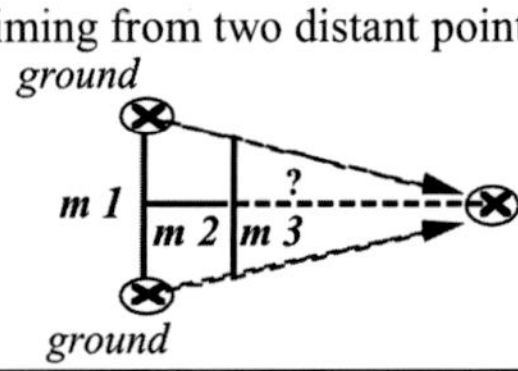	TWO SIMILAR TRIANGLES : *Two configurations* 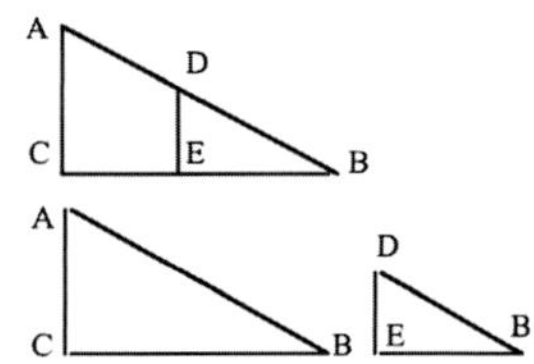 WITH THE HYPOTHESIS OF PARALLELISM: *Seven equalities between the ratios geometrically possible* 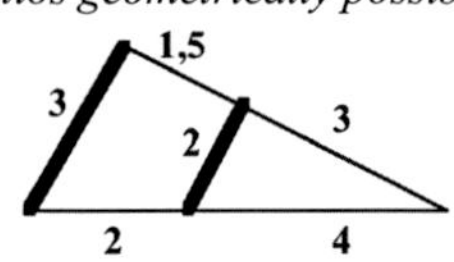

Fig. 3.12 The three real situations for applying Thales theorem

underlying mathematical activity and, hence, difficulties of deep incomprehension that the learning of mathematics raises.

Registers are semiotic systems that provides the means to create new knowledge. A semiotic system must meet two conditions to be a register. First, it should be able to produce representations that allow *to have access to perceptually or instrumentally inaccessible objects, and to explore all that is possible*. Then and above all, to open a field of specific operations that allow transforming the produced representations into new representations.

Of course, from a mathematical point of view, only the demonstrations allow to produce new knowledge. But from a cognitive point of view, we cannot understand the mathematical reasoning i.e., to acquire a minimum autonomy of initiative and control to carry out any mathematical activity, without becoming aware of the specific operations of each of the mobilized registers.

As an example of registers, we considered language and Euclidean geometric figures whose shapes can be recognized or constructed materially in 2D or 3D/3D. This choice may seem paradoxical because they are two culturally common registers. And, the semiotic revolution has marginalized them more and more and instead developed new kinds of semiotic representations to practice mathematics.

Algebra and analysis developed with the symbolic expressions of relations and with the representation systems based on the coordinates of points. Descartes did not dissociate them when using the term "register"to designate them. They have become dominant very quickly in mathematical activities, as they were immediately used in the upcoming physics[21] and allowed the discovery of new continents in mathematics. The development of the other sciences, based on the measurement and quantification of data collected experimentally found in these two mathematical registers the appropriate means to describe, analyze and modelling the observed phenomena. With these registers appeared a purely computational discursivity and the visualization of algebraic curves and functions. *This triggered a decline of the role of language in the development of scientific thought in favor of formalisms and diagrams, some of them being specific to a particular discipline*, as in chemistry.[22] A split then occurred between the natural language and scientific languages, i.e., between two heterogeneous kinds of rationality. It never ceases to increase, creating a complex and paradoxical situation for science education, and more radically for mathematics education. So why have we privileged the natural language and Euclidean geometric figures for introducing the concept of the register?

The language continues to be a necessary register in mathematics, if only to enunciate definitions, theorems, conjectures, etc. *An invisible divide line, but very deep, separates the use of language in mathematics from that in other fields of knowledge*. Whether to designate, deduct or enunciate, mathematics do not mobilize the same kind of discursive operations in both epistemological situations (above, Chap. 2, Fig. 2.4). The ways to designate, define, and reason in mathematics are strange, often contrary to those practiced spontaneously in an exchange or a debate outside mathematics. Similarly, when it comes to "seeing". The visualization of geometric shapes has nothing in common with the Gestalt recognition of shapes and objects in the images, or even with the graphs that visualize functions. This invisible divide is not taken into account in the teaching of mathematics. And that creates an insurmountable misunderstanding between teachers and students whenever natural language and geometric visualization are used in mathematical activity.

The operations specific to each register are cognitive operations. This means that the subject must have been already become aware of those ones to be able to perform them intentionally and spontaneously. This is the condition so the use of the mobilized register fulfills the three cognitive functions, which are involved in the ACT OF THINKING (*Noesis*).

- The production of representations that are representation of something, because thinking is always thinking of something. Unlike coding information, *the cognitive value of producing a representation in a register is determined by its reference to something*. And this production is about objects *that have been already*

[21] The development of these new means of representation (extensive to the negative coordinates while Descartes was limited to positive coordinates, infinity symbol etc.), allowing to use all the possibilities of representation and proper transformation of these registers, was partly conducted by Wallis, whose works (1655, 1656) paved the way for Newton.

[22] Granger, G. (1979). *Langages et épistémologie*. Paris: Klincksieck.

more or less explicated or shown in previous semiotic representations produced by others or by ourselves. There is, then, the introspective illusion that the actual thought of the object would precede the production of its semiotic representations. Mental representations are internalized semiotic representations.

- This production is first an objectification. *The production of semiotic representation precedes somehow the thought of objects* that are represented. Then, we become aware of what we were not aware before trying to produce for ourselves semiotic representations of these objects. This is evident with the use of natural language. The practice of a private discourse, a discourse for oneself, and intimate, empathic listening of its echo, is crucial for the individual development and in learning. But, it also works through the production of drawings, diagrams, or graphics.
- The transformation of representations either by conversion or by treatment. The production of semiotic representations in "intellectual" thinking, and more specifically in mathematical thinking, is more than a process of association of various representations related to the first produced by oneself. *The production of new representations only depends on the change of register and the substitution operation specific to the register selected.* Thus, treatments are the processes that generate new representations in the same register by one specific operation of substitution.

The mathematical thinking always mobilizes at least two registers. Even if, from a mathematical point of view, the mathematical work seems to use only a single register, the necessary understanding to run it requires the mobilization of a second register at least implicit, and their coordination. Furthermore, there are fields of mathematics wherein we explicitly mobilize at least two registers. Thus, in geometry, we mobilized the language and the visualization to deconstruct dimensionally the shapes perceptively recognized, while a third register is requested to calculate lengths, areas or ratios. Besides geometry, there is also the Analysis with the use of graphs and symbolic expressions of relations. All the problems that present real situations, such as additive and multiplicative problems, also mobilize three registers: the natural language, numerical expressions with the use of symbols of operation and, implicitly, diagrams.

The mobilization of a second register is required to discern and recognize which meaning units are mathematically relevant in the content of the representations produced in the first register! It is not enough; it is also necessary to coordinate the registers in such a way that they work in synergy. In other words, *it is not sufficient to juxtapose representations of different registers so that the students "see" the correspondences between the units of meaning mathematically relevant of the different juxtaposed representations.* It is why the conversion of the representations is the first threshold of mathematical understanding. It is in this kind of transformation that students can become aware of the representational functioning specific to each register.

The relationship between thought and transformation of semiotic representations is not at the level of a particular register of which we could analyze the cognitive functioning independently of the others. And even less, in any semiotic representation that we could consider independently of the other ones possible and their transformations. The term *"noesis"* was used by Aristotle and taken up by Husserl to describe the act of thinking. *Using the term "semiosis" to designate* the synergistic activation of at least two registers in the production and transformation of semiotic representations, we can say that there is no *noesis* without *semiosis*. Understanding in mathematics that first consists in recognizing the mathematical object represented and the possible transformations of their semiotic representations depend on the *semiosis*. As long as they cannot achieve this cognitive activity, students have a block in their learning of mathematics that is interpreted as 'incomprehension', 'not knowing how to do', 'loss of meaning', etc. *This underlying cognitive activity is irreducible to the objects and mathematical contents, which are taught.* Without its development, the teaching of mathematics cannot contribute in any way to the personal development of the mind. And the knowledge assessed as acquired is no longer recognized in the mathematical or real situations, in which they should be used.

The semiotic representations have a fundamental phenomenological property. They are *TRANSPARENT TO WHAT THEY REPRESENT when they work as semiotic representations for those who produce, understand or transform them.* This is the case, for example, when we speak or listen in our mother tongue. We understand straightaway what is said, without noticing the words spoken. This property is due to the nature of consciousness, which is first aware of what it aims at, i.e. immediately and fully in what is aimed and from the moment it aims at it. In other terms, as Husserl never ceased to repeat it, the achievement of the act of aiming at anything is transparent to the modes of grasping what is aimed.

This phenomenological property of semiotic representations is always used as an objection against the registers. Thus, mathematicians claim they have an immediate access to mathematical objects (they "exist") and only secondarily be aware of their semiotic representations. Moreover, interest in the semiotic representation registers would introduce an intermediary, and, therefore, a screen between the mathematical objects, their properties, and the mathematical understanding. How many times have I not heard this objection from the didacticians! And from this transparency of semiotic representations that the only experts can subjectively experience, they conclude the non-semiotic character of the cognitive functioning of the mathematical thinking.

The facts are quite different. *Semiotic representations are only transparent when there is immediate and spontaneous recognition of what they represent.* Also, attention can always switch what it currently focuses on. Thus, consciousness can instantly change from the mathematical object aimed to the semiotic representation used to get access to it, but without noticing this reversal. Transparency of semiotic representations does not mean their absence. Rather, the teachers are continuously confronted with the non-transparency of semiotic representation for the students. And, faced with the variety and complexity of semiotic representations whatever the register used, the students cannot control anything if the conversions to perform are not congruent and if the treatments to be made are not an automatic calculation algorithms. It is as if the students did not have any ability to recognize what is represented, not even the first transformations that they could perform to tackle on their own any problem even simple.

Chapter 4
The Registers: Method of Analysis and Identification of Cognitive Variables

We need to be able to accomplish spontaneously two things when facing a mathematical production, whether it is given as a result, explanation, proof, or requested as a problem to solve or even an activity to be executed. Firstly, we should recognize the meaning units, that is, the mathematically relevant data or information *in the content of representation given*. Secondly we should begin to transform these meaning units either by converting them into another register or by *performing the treatment operations specific of the register selected. These are the two cognitive conditions required* to understand and do anything in mathematics. They concern the specific way of working and thinking in mathematics. Otherwise, most of the students can only feel unable to learn mathematics.

We had to introduce the notion of the register of the semiotic representation to describe the cognitive functioning underlying the act of thinking mathematically (*noesis*). All mathematical productions are necessarily semiotic from an epistemological perspective. Also, analysis of the various mathematical activities, either proof processes, or solving problem based on data from real situations, consists of transforming the semiotic representations. And what is commonly called "conceptualization" in mathematics is based on the *cognitive coordination* between the use of quite different semiotic systems for producing new representation.

The interest of modeling the cognitive functioning of thought in terms of registers is not theoretical, but first methodological. It provides the tools to analyze the two conditions required to understand and do mathematics by oneself. The first is the coordination between two registers whose development gives the ability to convert any representation given from its register A into another of the same object in another register B. The second is the internalization of the intrinsic operations of each of the two registers. The principle of the analysis is as follows: to understand what a semiotic representation A_1 represents and how it represents it, one must consider a second representation B_1 that refers to the same object, and make systematically change its content, in order to see whether anything change in the content of the first representation A_1. In other words, we can never discriminate what is math-

R. Duval, *Understanding the Mathematical Way of Thinking – The Registers of Semiotic Representations*, DOI 10.1007/978-3-319-56910-9_4

ematically relevant in any representation A_1 if we consider it alone, regardless of all those in which it can be converted.

The first condition leads to the following procedure. Firstly, generate all the modifications of the content of a given representation A_1 that are striking changes, whether or not the object represented changes. Secondly compare them with the parallel changes in the initial representation B_1 referentially equivalent to A_1. We can then see which changes of A_1 lead to changes in the content of the referentially equivalent representations in the register B, and which ones result in no change. Thus, the second register B reveals the mathematically relevant meaning units in the representations of the initial register A. The use of a second register is methodologically important. It allows recognizing why and how two representations, which are produced in two different registers, are or are not representations of the same object. Or, conversely, if two representations, which are produced in the same register, and of which the contents look almost the same, are or are not representations of the same object.

Regarding the second condition, the procedure is limited to a single register without mobilizing, even implicitly, another register. Then, it is necessary to make an inventory of possible operations which change directly either some part or the global organization of the content of A_1 into another content A_2 that is both derivate from A_1 and new compared to it. Naturally, this is done without worrying whether or not the represented objects are still the same. This inventory can also be conducted on a wide range of mathematical practices and activities. We can then identify the processing operations that are possible within a register and particular to it. So, we have been able to identify the purely figural transformations of geometrical configurations in which the heuristic power of mathematical visualization consists (above, Chap. 3, Sect. 3.2.2).

The registers are the tools that allow us to analyze all mathematical production, and primarily those devised to teach.

To show this, we will begin by analyzing the case of the simplest productions, the Cartesian graphs. This example presents a twofold interest. Firstly, it shows the cognitive difference between code and register. In education, these representations are introduced as coding technique, while being used as a representation register. Their visualization is qualitative, and the understanding of what they represent requires their cognitive coordination with the register of symbolic expression of relations. Secondly, it shows the kind of representation we want to focus on, for didactic purposes, had to be analysed by taking in account the couple of these two registers.

However, an overview of all the couple of registers that mathematical activity can mobilize is indispensable to analyze mathematical activity by using these tools. The conversions and treatments, which are required to solve problems, can depend on quite different couples of registers according to the problems to be solved. It is why a classification or registers is essential to use the registers as tools. Such a classification is needed to analyze the implicit or lateral representations cognitively required by mathematical activity. It also allows identifying all the determining cognitive factors in developing the understanding in mathematics and transfer mathematical knowledge to situations different from those in it was introduced. To illustrate these points, we consider a basic and well-known example, the additive problems.

In the production of semiotic representations, one should never confuse the mobilized register and *the phenomenological mode in which semiotical representations are produced.* As the representation of a mathematical object can be produced in different registers, also the production of its representation can be made according to different phenomenological modes. Here, we have two independent variation sources. Mathematically, the choice of the phenomenological mode of production is neutral, but not for learning. Thus, speaking and writing are both required so that students can become aware of the operations intrinsic to each register. The issue of the contribution of computers to the first mathematical learning should be examined from this perspective.

Didactic research has always singled out the observation of students in classrooms. That has raised the complex problem of analysis of student's productions, because they are very often far removed from what is expected as a mathematical solution. Hence the need for a triple analysis. A mathematical analysis in terms of right answer and validity of its justification. A cognitive analysis of acquisition of knowledge in terms of degree of autonomy and margin of progression. And, a diagnostic analysis of the reasons that led to success and acquisition or to failures and mental blocks, i.e. an analysis according to the factors that were taken into account or not taken in the organization of the teaching. Here we will limit ourselves to the organization of sequences of learning activities and analyze as an example geometric activities for 10 year-old students.

4.1 How to Isolate and Recognize the Meaning Units Mathematically Relevant in the Content of a Representation?

The case of graphs is typical since they are the simplest possible representations to construct. There is only one coding rule to be applied: a couple of numbers corresponds to a point on a plane plotted according two oriented, graduated axes. Unlike the geometrical figures, they seem easy to use: we recognize immediately a line or curve, whether it increases or decreases, and in the case of curves, we see at first glance the number of branches, the inflection points, maximum and minimum points. They belong to the general culture because these representations are used in all fields of knowledge. They allow visualizing the *quantitative variations of phenomena due to a factor*: e.g., daily temperatures depending on the seasons and over the years.

4.1.1 Production of Graphs and Their Equivocal Obviousness

The encoding rule is insufficient to represent a linear or quadratic function. *It allows only to plot intersection points on a grid reference,* AND NOTHING ELSE. It is necessary to perform two other operations to obtain a representation, as seen in the sequence below (Fig. 4.1).

The operation of joining the plotted points is a transgression of the coding rule, because we replace two plotted points (figural units 0D) by the *VISUAL CONTINUUM of a stroke* (figural unit 1D), *which takes several qualitative visual values*: it is right or it is curved, its orientation is upwards or downwards, it crosses the two axes or only one, etc. Certainly, we can justify this operation by invoking another kind of representation, as the table of values or a definition of a property of the traced curve. Nevertheless, in this case, we renounce analyzing the specific functioning of the graphical representations and make the students become aware of it.

The operation of plotting intermediary points leads to another chain of smaller strokes. If they still appear on the same line we can stop and draw the line. If they do not appear on the same line, we have to repeat the operation until the chain segments coincide visually with a nice, smooth curve. *Here we play with the threshold of visual discrimination of the human eye.* And then a chain of very small segments is seen as a perfect curve.

From the only coding rule, *which can never lead beyond* series *of points*, to these two operations there is a DIMENSIONAL JUMP IN THE VISUAL CONTINUUM of lines and curves with their qualitative visual values. It is because of this dimensional jump that Cartesian graphs become a semiotic system producer of a new kind of representations, and not because of the coding rule that assigns numerical values (abscissa and ordinate) to each point. It enables to make on the visual continuum, operations that are not possible with the points.

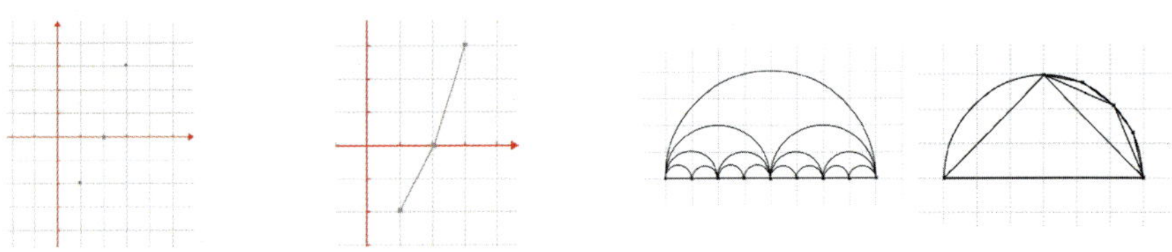

1. The coding rule allows only to plot intersection-points on the grid reference.

2. We JOIN THEM, *and we get a chain of segments.*

3. We can ENCODE INTERMEDIATE POINTS. This is the same as zooming in on an interval, considering a finer grid reference. *Each segment can thus be transformed into a new chain of smaller segments.*

A. How many times is it necessary to reiterate this division so that the *semicircles are visually confused with the diameter* of the starting large circumference?

A'. How many times is it necessary to divide the segments *so that the chain of segments coincides visually with the curve?*

Fig. 4.1 The three semiotic operations to produce graphs

This jump from discrete series of points to the visual continuum of lines and curves creates obviousness, which is cognitively equivocal. Because no segment can approach the arc with which it has common extremities, what we see may be deceptive about the mathematical propriety of continuity of the object represented. This problem is caused by the use of graphing calculators. We believe we see on the screen what the software did not draw. Thus the introduction of both Cartesian representations and functions in Mathematics Education pass over the greater enigma of the impossible relation between a segment and an arc, which is the challenge of approaching a curve by a sequence of segments.

Replacing a sequence of ever more closer points by the 1D visual continuum raises a difficulty, which is independent of the fact that we use grid paper or a graphing calculator. *Whatever the graphical representation plotted or given (line or curve), how to recognize the qualitative visual values, which are mathematically relevant?* These qualitative visual values are the meaning units of the register of graphical representations. Without their spontaneous recognition, the use of graphical representation is useless or misleading.

4.1.2 Methodology to Isolate the Mathematically Relevant Meaning Units in Any Representation Content

We cannot distinguish the mathematically relevant meaning units in the content of a representation *without converting implicitly or explicitly the representation into another register*. However, a single conversion is not enough either to recognize the relevant meaning units or to justify the objective relevance of their one-to-one mapping with the relevant meaning units of a representation referentially equivalent in another register. The content of the starting representation need to be varied in a systematic way and each variation of content had to be compared with that of the equivalent representation in another register. It is this methodology that allows discriminating the relevant meaning units, which are merged into the whole content of a semiotic representation. We have applied it for the first time to graphs in a work published in 1988.[1] This work highlighted two phenomena: the importance of the conversion for not confusing the object represented with the content of the representation, which depends on the selected register, and the profound difficulties that conversion caused systematically for the great majority of students.

The often misunderstood, crucial point of the methodology is the systematic variation of the representations from which we want to isolate the mathematically relevant units of meaning. For Cartesian representations, *these variations should be purely visual and must correspond to the visual qualitative oppositions in the visual recognition* of the graph form, its orientation, and its position relative to the axes. Thus, for all the linear graphs, the straight line (or segments of lines), we have the following five visual qualitative oppositions:

[1] Duval, R. (1988). Graphiques et equations: l'articulation de deux registres. *Annales de Didactique et de Sciences Cognitives*, n. 1, p. 235–255.

- The line goes up or down according to the orientation of the y-axis,
- The line is either closer to the x-axis or y-axis according to the symmetric division of the plane into quadrants by the coordinate axes,
- The line passes or not through the origin,
- If it does not pass through the origin, it passes either above or below,
- The line is either parallel or not to one of the axis.

Therefore, for each of these five visual oppositions, we get two visual values according the Saussure semiotic criterion. They constitute the meaning units possible of linear graphs. Thus, each line plotted on a Cartesian plane has necessarily five visual values. Nevertheless, these values tend to merge into figural units 1D of lines or curves over the background of two axes. Very often one seems to impose itself. How then can we isolate each of these visual values and recognize those mathematically relevant and why?

It is the first requirement in experimental sciences: how to separate the different factors that interfere with the creation of a phenomenon, and can we separate them? The fundamental methodological rule is to vary each factor while keeping all other constant. Here, the factors are visual qualitative oppositions that we have just indicated. We can then construct the variation table of the visual content of linear graphs below (Fig. 4.2)

The visualization power of linear graphs lies on the spontaneous discrimination of all these qualitative visual values. Every line plotted according two axes has necessarily two or three of these qualitative values, and it is needed to recognize them to see what any linear graph *shows*. That means that students must already have acquired the ability to discriminate the visual values of which the combination distinguishes any linear graph from the other possible graphs, *regardless of the numerical values of the coordinates, coefficients, and constants.*

The conversion of that variation of linear graphs into equations allows isolating what changes in the writing of the equation *when one visual value is changed while the other visual value remain constant.* It is this almost experimental observation task,—variation of the visual value of graphs and covariation of the categorical values of equations—which enables us to become aware of what is mathematically relevant in the visual content of the graphs. The table below shows the correspondence between the meaning units of the visual content of graphical representations and the symbolic content of equations. It also shows that the same methodology can be used, taking the equations instead of the graphs as the starting register. The inverse conversion of equations into graphs, without any calculation of the coordinates, is a cognitive quite different task. Both conversions, direct and inverse, are the test for the cognitive coordination of the two registers (Fig. 4.3).

This method allows us to analyze all graphic representations (Duval, 2003). We have just to modify the shape of 1D figural unit, for example, considering the curves and not the lines, or the dimension of figural units, for example, surfaces (2D figural units). Then the visual changes will correspond to the variations of the degree of the variable letter, to the fact that the relationship is an equality or inequality. We can also vary the axis grading system.

1. *THROUGH* the ORIGIN

   And *symmetrically divides* the quadrants (not closer to any particular axis).
2. *THROUGH* the ORIGIN

   And *goes up*, being closer to one particular axis (not dividing symmetrically the quadrants).
3. *THROUGH* the ORIGIN

   And *goes down, becoming closer* to a particular axis than the other
4. *DOES NOT THROUGH* the origin

   And *goes up*, parallel to the symmetric division of the quadrants.
5. *DOES NOT THROUGH* the origin

   And *goes down*, parallel to the symmetric division of the quadrants.
6. *ABOVE* the origin

   And *goes up, becoming closer to* one axis than the other
7. *ABOVE* the origin

   And *goes down, becoming further from* one axis than the other
8. *BELOW* the origin

   And *goes up, becoming closer* to one axis.
9. *BELOW* the origin

   And *goes down, becoming further from* one axis.
10. *PARALLEL To* one of the axes.

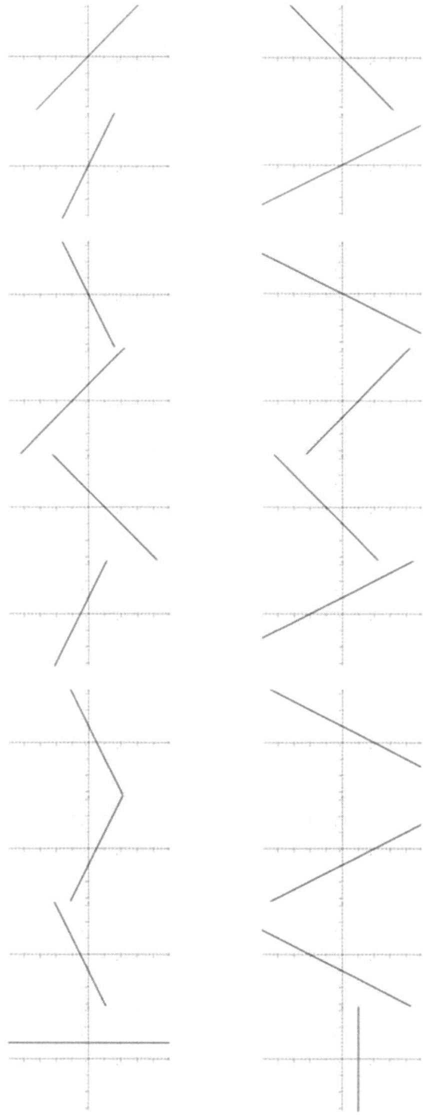

Fig. 4.2 Variations of the oppositions of the visual values on a linear graph

The cognitive efficiency of the conversion, used as an analysis tool of what is mathematically relevant in the content of any semiotic representation depends on how we change the representation to be converted. There is here a theoretical and methodological condition. *Changes must be made according to the meaning units, which are specific to the register of the representations to be converted* - here the visual units 1D, 2D—*and not according to the represented mathematical objects.* For example, it does not matter that in the four graphs below (Fig. 4.4), one does not represent a function,

Visual qualitative values of a linear graph			Equations
Tilt direction	**Angle formed with the *x-axis***	**Position on the *y-axis***	**Example**
Straight line going up **(+) X**	Symmetric division coefficient **= 1**	passes through the origin **+0** passes above the origin **+1** passes below the origin **–1**	y = (**+1**) x y = x + 1 y = x – 1
	Bigger angle coefficient **>1**	passes through the origin **+0** passes above the origin **+1** passes below the origin **–1**	y = 2 x y = 2 x + 1 y = 2 x – 1
	Smaller angle coefficient **<1**	passes through the origin **+0** passes above the origin **+1** passes below the origin **–1**	y = 1/2 x y = 1/2 x + 1 y = 1/2 x – 1
Straight line going down **–X**	....	....	y = – ....

Fig. 4.3 Symbolic expressions of the oppositions of the visual values of a linear graph

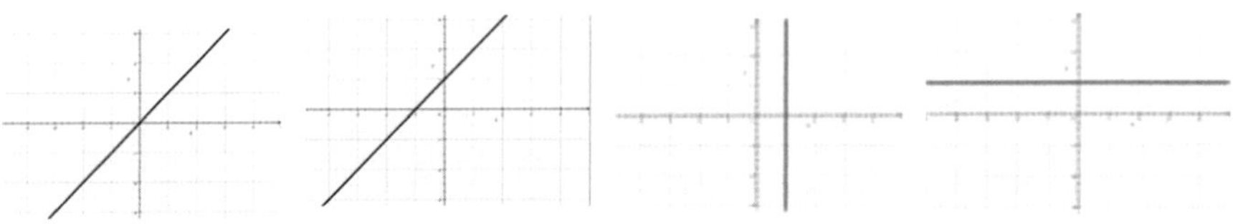

Fig. 4.4 The articulation between graphs and equations is independent of the object represented

another represents a linear function and the other two, affine functions. The objective is to recognize immediately which variations of the equation correspond to variations of the graphs. *Without this recognition, which we call mathematical literacy,* we cannot have any articulation, no comings and goings between the graphical representations and algebraic writings.

After using this methodology for the first time, further work was undertaken to analyze among the curves, the ones that are graphs of quadratic functions, referring to an analysis in terms of registers. However, in many cases, the task of conversion is organized without really considering the condition of systematic variation of the representations in the starting registers. One might therefore question the validity of the analysis and the significance of the results. Similarly, the software is now s an immediate and very varied production tool according to the register of algebraic expressions. Nevertheless, its didactic use often raises the same question, as we shall see (Sect. 4.4).

To subordinate the introduction of graphs to this of the notion of function, as it is always done in the curricums and in didactical sequences, and, more generally, to introduce different forms of graphs based only on the mathematical objects they represent, is to engage the students in a double impasse. First, we neglect the importance of mathematics literacy, even in the visualization process. Second, we forget that the notion of function requires a third register, the verbal statements that characterize the correspondence criteria between a point and its image and that allow to define it. Therefore, *it is reduced to a purely numeric input in the graphical representations,* and we assume that the students spontaneously coordinate by themselves proportionality tables and geometric visualization of similar triangles (see Annex).

4.1.3 What Kind of Task for Developing the Recognition of Mathematically Relevant Meaning Units?

This method of analysis is based directly on the special epistemological status of the mathematical objects and the particular cognitive functioning that their special status requires developing to "do" math. Does it allow analyzing the recurring incomprehension problems caused by learning mathematics and highlighting the factors that favor the development of understanding? In other words, what kind of task allows to extend an analysis of mathematical productions into an analysis of the underlying processes of recognition of what is mathematically relevant in any semiotic representations?

From the mathematical point of view, the learning situation focuses on problem-solving. From a cognitive point of view, let us first focus on the organization of recognition tasks. Two deep differences separate the problem-solving from the recognition tasks. First, the problems are chosen according to the objects and mathematical properties to be worked out. On the contrary, the recognition tasks are chosen based on the cognitive variables that can be identified by the method of analysis of mathematical productions we have just presented. Furthermore, the cognitive activity required by solving problems is global, what involves an implicit part that can vary considerably, and which may remain incomplete or break down. On the contrary, in a recognition task the activity is focused on the crucial cognitive operation that is implicitly required to solve any problem, and each student can entirely check its achievement for him/herself. The task requested is simply to choose an answer from several possible ones. Also, each choice should not require any time. The success of a recognition task should be spontaneous, or at least not take longer than a minute or two. Otherwise, we cannot talk of acquisition even if students end up performing some of the items of the sequence of recognition tasks. It is the margin of spontaneous recognition that opens up or closes the door to new learning. Indeed, the initiative and control capability when solving a problem depends entirely on the recognition of what is mathematically significant in any problem dataset, i.e., the ability to differentiate the variations of representations of the dataset in one or more registers.

Therefore, in 1988, we proposed multiple-choice tasks in the first organized observation work according to the method of analysis of mathematical productions. Only three visual oppositions were retained among the 20 possible situations. One equation had to be chosen for each plot from the four proposed. After teaching about the affine

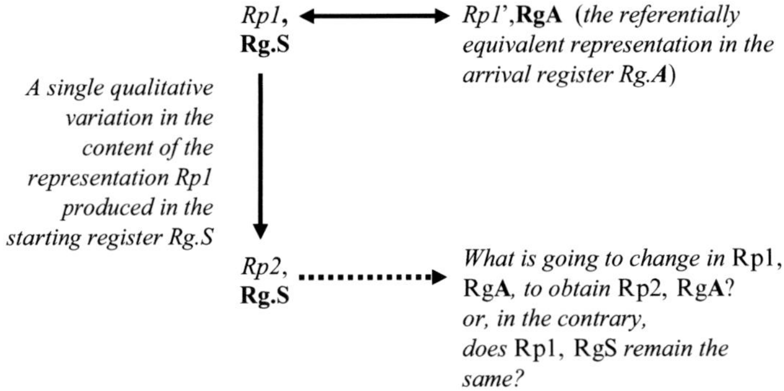

Fig. 4.5 Scheme to organize an elementary task of recognition

functions, only two-thirds of 15 to 16-year-old students were able to discriminate the graphs of the equations y = x and y = −x. This ratio dropped to one-fourth when the students had to associate the graphs to the corresponding equations, y = 2x; y = x + 2; and y = (1/2)x. *That required recognizing the relevant visual values of a straight line on a Cartesian plane and nothing else*. We could have asked the students to produce the different equations since they appear elementary after being taught functions. However, this could have prompted to use calculation procedures with no results, which was confirmed later by some individual observations. Graphs allow neither to see nor understand, nor anticipate what the equations express if you have not learned how to recognize the qualitative variations of the any plotted 1D visual continuum, and relate them to the variations of some symbolic terms of algebraic writing. And, conversely, starting from the variations in the algebraic writing of equations.

The cognitive analysis method associated with recognition tasks is an observation tool that allows highlighting the phenomena we cannot observe solely through the classroom observations. It also provides a scheme to organize the learning situations in order to recognize the mathematically relevant MEANING UNITS in the graphical representations, problem statements, literal writing of numerical relations, symbolic expressions of relations, and geometrical configurations. Therefore, it is possible to represent an elementary cognitive task of recognition in the following scheme, where *Rp* is a representation of an object, RgS and RgA are the starting and arrival registers in which the representation of an object is produced (Fig. 4.5).

Clearly, the organization of an experimental observation activity of this kind requires that we preliminarily identify the variables appropriate to the register whose we want to make the students becoming aware of the specific way it represents mathematical data, operations or relations. A table of variations of oppositions of visual values allows making this organization for the graphs. Then, everything depends on the degree of freedom given to the students to observe and describe algebraically or verbally the observed visual variations. In any case, this kind of activity is diametrically opposite to problem-solving. But it is the cognitive condition to become able to solve problems, because solving problem always involves several cognitive factors, which cannot be separated to help students understanding the mathematical way of working out any problem.

The choice of the second register RgA is crucial because it should help to reveal the mathematically meaning units in any representation content produced in the first register RgS. It was obvious in the case of the qualitative visual values of graphs. RgA must be the symbolic writing of relations. Moreover, to analyze this register, the reverse choice imposes itself naturally, even though this is often not done in teaching where the coding rule to convert the equations or inequalities into graphs is only put into practice. For other registers, those that are not specific to mathematics, the question of choosing the second register is more acute and complex as we shall see.

4.2 The Analysis of Mathematical Activity Based on the Couples of Mobilized Registers

From a mathematical point of view, a single register is sufficient, because the only transformations epistemologically important of the semiotic representations are the treatments. Mathematical justifications and proofs are always based on this kind of transformations. The fact that we do not always use the same register, according to the mathematical objects we work with, does not change that. The treatments are the intrinsic transformations specific to the selected register. Thus, it is not surprising that most didacticians and teachers seek directly to analyze these intrinsic transformations according to the properties of the represented objects, and not indirectly in relation to a second register that would need to be mobilized to understand the intrinsic transformations specific to the selected register.

On the contrary, the analysis of the cognitive functioning of thought required by mathematics shows the need for a simultaneous and coordinated mobilization of several registers to be able to understand mathematics and solve problems. The real mathematical activity is never limited to using a single register. It is always wider and more complex than the explicit productions in the register where the treatments are carried out. A second register, or even a third one, had to be mobilized to find out the treatments to be done, and therefore, to control the treatments made in the selected register. This mobilization may be done explicitly for the parallel production of a second register such as in geometry or analysis. Nevertheless, it remains usually implicit. *In other words, in mathematics, we never think in a single register, but several at once, even if the explicit productions favor a single register.* Also, this requires a ceaseless activity of conversions, which are implicit, but should be more or less spontaneous.

Therefore, it is necessary to take into account all the registers used in mathematics to analyze the mathematical activities organized with the purpose of learning or training. To analyze problem-solving, we cannot only focus on the register where the mathematical treatments give the solution to the problem posed. The mobilization of other registers related to the data set and the way they are represented are also essential. Here, we find back the condition of applying the general method of analysis presented above, i.e. the possibility of considering other possible couples of registers.

We will present a classification rather than enumerate all registers used in math—enumeration that would seem insufficient and, above all, would not explain how their mobilization is intrinsic to the cognitive functioning of thought. The interest is both theoretical and methodological. It situates all transformations of representations that constitute the mathematical activity, namely, the operations that are proper to each register and the various kinds of conversion. It allows evaluating *the cognitive distance* between the representations of two different registers, which can change drastically according to the couples of registers. This classification of registers of semiotic representations is radically different from that of Pierce (Chap. 1, Sect. 1.3.2). It relies solely on four fundamental characteristics, which in twos are mutually exclusive.

Registers are discursive or non-discursive. Of course, among the discursive registers, no one considers that the natural language falls into the same kind of representation as formal languages or the symbolic writings. Similarly, among the non-discursive registers, geometric configurations do not fall into the same kind of representation as Cartesian graphs, diagrams or hand-drawn images. Each of these characteristics determines specific means of representation and transformations of representation (Chap. 3). Thus, each register combines one of these two mutually exclusive characteristics. *This classification leads to distinguish two quite different kinds of registers, from a mathematical point of view and from an educational point of view.*

Registers are multifunctional or monofunctional. The monofunctional registers are specific for mathematics. Multifunctional registers fulfil outside mathematics the communication and objectivation functions, and under particular conditions the treatment function. The dividing line is clear in mathematics. But it is not anymore for teaching primary or secondary school students, because multifunctional registers are at least as important as the monofunctional. The main trouble is with the natural language. It is used to fulfil the communication function between students and teachers and to fulfil the mathematical treatment function whenever it is needed to justify or prove. These two practices of the natural language are in conflict with each other.

We obtain, then, the following table of the registers used in mathematics (Fig. 4.6).

We could now complete this table by indicating with arrows the two kinds of transformation: conversion and treatment. An arrow between two cells corresponds to the conversion of a representation from a starting register to the arrival register. However, it would necessary to differentiate direct and inverse conversions between two registers. That results in the most general and recurrent block in the learning of mathematics, but rarely taken into account in the educational studies and in classrooms. Success in converting representations in one sense does never involve the ability to recognize and do the inverse conversion. Direct and inverse conversions are two cognitive tasks as different as going up or down a steep path on a mountain. Therefore we have to mark the inverse conversion. In other words, the synergistic coordination of a couple of registers requires the ability to convert the representations in both directions and not in only one. Now, if we wanted to bring up the treatments in the table, it would necessary to insert a spiral arrow inside each of the four

	DISCURSIVE Registers *Linearity based on succession* to produce and organize sequences of words or symbols	NON-DISCURSIVE Registers *Simultaneous apprehension of a two-dimensional organization of nD figural units*
MULTIFUNCTIONAL registers : *the transformations of the expressions are NON-ALGORITHMIC*	**NATURAL LANGUAGES.** *Three hierarchically included operations* (naming the objects, enunciation, and reasoning) *with their corresponding meaning units* *Two phenomenological modes of production:* oral or written	**ICONIC : IMAGES** *freehand production and internal conservation of topological relations characteristic of parts of the object.* **NON-ICONIC : GEOMETRICAL FIGURES.** *Three independent operations:* instrumental construction, mereological reconfiguration and deconstruction of the 2D dimensional shapes.
	(Auxiliary transitional representations *for free external operations*)	
MONOFUNCTIONAL: Registers: *the transformations of the expressions are ALGORITHMIC*	**THE SYMBOLIC WRITING** *for an unlimited substitution operation:* numbering system, algebraic writing, formal languages) *A single phenomenological mode of production:* writing	**CARTESIAN GRAPHS, DIAGRAMS** *strokes and arrow joining marks or nodes* For the graphs :operations of zooming, interpolation, change of axis.

Fig. 4.6 Classification of types of semiotic registers

cells determined by the intersection of margins.[2] We get something like the result of an MRI scan of semiocognitive areas that must be mobilized in a mathematical activity (see Annex).

This classification is a tool for analyzing the registers used explicitly, and those implicitly mobilized, but not considered in the mathematical analysis of problem-solving. It also allows to clarify the cognitive functions that they fulfill. An analysis made only in relation to mathematically represented objects confuses or hides the real cognitive activity that should be performed to solve the problem. Thus, for geometric figures, we talk about the properties of the figure using words that usually imply a dimensional deconstruction from what the figure allows us to see. Then, we use the properties stated in definitions to deduce and see something else, without dealing with the 2D shapes recognized in the figure. For graphs, we use the knowledge originating from another register to cover the discontinuity of the points in a visual continuum because we know that it is such a function and not another. In the classroom, there is a considerable semiocognitve distance between what the teacher writes on the blackboard, and what he tells orally to explain what he is writing. All

[2] Duval, R. (2006). The cognitive analysis of problems of comprehension in the learning of mathematics. In: Saenz-Ludlow A., Presmeg N. (Eds.). *Semiotic perspectives on epistemology and teaching and learning of mathematics. Special issue. Educational Studies in Mathematics,* 61, p. 103–131.

that requires that the students perform quickly and implicitly a lot of various conversions.

4.2.1 *Congruence and Non-Congruence Phenomena in the Conversion of the Representations*

To change the register of a given or obtained representation after a very elementary treatment is the first step of mathematical thinking. Without this step that should be more or less a reflex, no mathematical reasoning or activity is possible, we become blocked, we see nothing and we do not know what to do. Moreover, even if someone suggest how to convert the representation given, the misunderstanding remains. The key question remains unanswered: what would and will allow, in another situation, to recognize the relevant meaning units in the representation given? Certainly, these negative experiences are forgotten quickly, but they unconsciously accumulate in the course of schooling.

Two points are essential as regards the conversions of representations. First, *the conversions to be performed can be implicit or explicit, because they are always required in one of two following situations of multi-representation or mono-representation* (Fig. 4.7).

Although mathematically it is the same problem, there is a considerable distance between the two statements, I-I′ and II-II′, of the multi-representational situation. In the first statement, I-I′, the 2D figural units designated by the words "parallelograms" correspond to the immediately recognized 2D shapes: either in plane two 2D figural units partially one above one another or in the space the two sides of a roof or an open book upside down. In the second, II-II′, the correspondence is no longer there because the

MULTI-REPRESENTATION SITUATION *double entry: focusing back and forth on the two semiotic representations given*		MONO-REPRESENTATION SITUATION: *The semiotic representation given had to be converted into (diagram ?) and symbolic expressions of relations*
I ABED and BCED are parallelograms Prove that B is the middle point of AC	**I'** 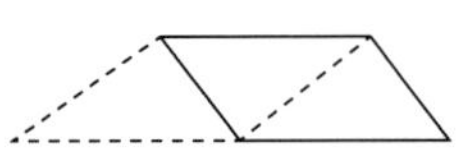	**III** A basin is fed by two sources, A and B. We let A flow for 4 hours and B for 2 hours, we obtained 64 liters. We let A flow for 3 hours, B for 4 hours, we obtained 62 liters. What is the flow of each source?
II AC and Je are parallel AB and IE are parallel IJ and CB are parallel Prove that E is the middle point of CB	**II'** 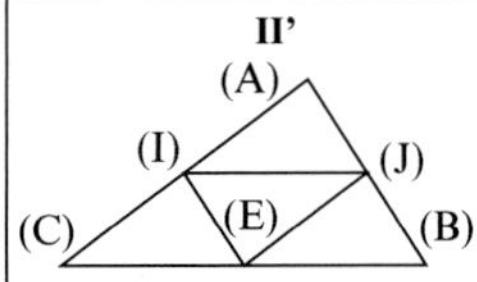	**III** A moped goes up a hill at 15 meters per second; then it goes down the other side at a speed of 21 meters per second. The total journey lasts 270 seconds and the climb is 126 meters longer than the descent. How long did it last the uphill journey?

Fig. 4.7 Congruence and non-congruence in the conversion of representations

statement refers to figural units 1D and the perception imposes the recognition of juxtaposed triangles. However, the perception always imposes the recognition of 2D shapes against 1D shapes (Chap. 3, Fig. 3.5). At the end of the school year, 15-year-old students had a success rate higher than 50% for the first problem but failed completely in the second. Also, among those who saw the second problem immediately after the first, few recognized the same problem. The success rate was very low.

There is also a considerable gap between the two statements, III and IV, of the mono-representation situation. The dataset of the verbal description had to be first converted into equations. In the first statement, III, there is an implicit double correspondence: between the two conditional statements and the equation to be written; and, between the terms of each sentence and those of each equation. We can say that this statement is transparent: we can almost see through the statement the system of equations that need to be written! This is not the case for the second statement, IV, which was very difficult even to 17 to 18-year-old students. This variation of congruence or non-congruence in mono-representation situation does not concern only the task of putting into equation the dataset of problem, but it also concerns the additives problems that are, actually, problems of arithmetic equations. The variation of congruence and non-congruence is one of the major causes of misunderstanding or misinterpretation of problem statements by the students.

Either in multi- representations situation or mono-representation situation, it is always the same fundamental cognitive process of conversion of representations that is requested. And, it is always the same variation of congruence and non-congruence between the respective contents of the starting representation and arrival representation *that facilitates or inhibits the conversion.* And it is independent of the mathematical relations represented. This variation goes against two ideas that are largely dominant ideas both in the educational theories and the teaching practices.

The first idea is that the multi-representation, i.e., the presentation of statements in parallel with diagrams, tables, images, would facilitate understanding. This is true outside mathematics, even though in many cases there is no congruence between the texts, diagrams and tables presented on the same page, but not in mathematics! Multi-representation is only an opaque juxtaposition for all those who have not yet become aware of the specific way of thinking and working in mathematics (Chap. 2, Fig. 2.2). Then, the question is not what kind of non-discursive register to choose for helping them, but how to make them able to convert the representation of any object from one register into another register. Mathematical activity begins with the synergy of two registers, whatever the registers chosen.

The second idea is that the mathematical problems should not be presented as mono-representation statements, but based on either a concrete or game situations that the students could work out more easily. And, so we criticize the *"word problems"*.

But, emphasizing the concrete context of the dataset just shifts the cognitive difficulty of the problem without helping understand how to solve it in other contexts. We have seen that simple numerical description of token configurations in squares could become not congruent in two moves of the encircling procedure and pose an extraordinary difficulty of geometric reconfiguration, unsurmountable for students (Chap. 2, Figs. 2.7, 2.8, and 2.9). In general, the use of materials, or non-discursive representations, raises the same troubles of non-congruence because the mathematical activity

The set of points whose abscissa **AND** ordinate have the ***SAME SIGN***	x (×) y > **0**	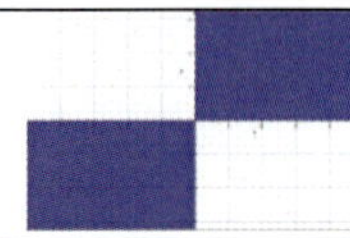
	The product of abscissa and ordinate is **greater than 0**	**Two** of the four **quadrants** determined by the abscissa and ordinate axes

Fig. 4.8 Comparison of three non-congruent representations

requires the mobilization of a second register of semiotic representation. Variations of congruence and non-congruence are there, between the material and the semiotic register to be mobilized.

When we analyse the problems chosen by the teachers, we observe the tendency to choose preferably the problems in which the conversions to perform are congruent while keeping the non-congruent problems for research sessions with more advanced students. That is only a tendency, because they do not have this didactic variable explicitly in their mind, and they often propose problems that require non-congruent conversions to students who already have difficulties. In reality, it is the practice of presentation and the use of problem-solving that should be questioned again. Instead of presenting each problem by itself, independently of the others, it is a whole range of possible presentations, organized according to the variations of congruence and non-congruence that the students should work, in order to discuss them and become aware of these variations.

Now, the second essential point as regards the conversions is that *their difficulty reflects the cognitive distance that separates the representation contents of the same object within different registers.* Consider the following three representations. Do their contents have meaning units, which can be easily mapped one-to-one in such a way that they could be recognized as representations of the same thing? (Fig. 4.8)

The answer is obviously not. The only meaning units that can be mapped one-to-one concern the words, the symbols and axes of "abscissa" and "ordinate". There is no congruence between the expressions "same sign" and "**>0**". For this last expression, the congruent wording is "greater than 0". Likewise, we can wonder about the congruency of mapping one-to-one the conjunction "and" and the multiplicative operation "**×**". In fact, it is necessary to reword in logical and operative symbols the word whose meaning is different in natural language: reunion becomes intersection. Similarly, to recognize the quadrants that are 2D figural units, it is necessary to break with the reflex act of researching the intersection points (0D figural units) and consider the terms "x" and "y" as corresponding to the 1D figural units of the four semi-axes. We observe that there is a lack of a strong criterion of congruence, i.e. the possibility of a one-to-one mapping direct and regular at the same time.

More generally, the cognitive distance between the discursive and non-discursive registers is always bigger than it looks. *There is an asymmetry between the two opposite sense of the conversions, due to the radical heterogeneity of the content organization rules, which characterize each of these two kinds of registers.* Let us return to the example of the only verbal designation operation of figural units in a geometric figure (Chap. 3, Sect. 3.1.2) (Fig. 4.9).

LANGUAGE ⟶ FIGURE For a verbal designation ***Uniqueness condition:*** *therefore a single figural unit*	FIGURE ⟶ LANGUAGE For a figural unit ***at least two different verbal designations,*** because it is the condition to ensure the continuity of the discourse between two successive propositions
1. Draw a **segment** [AC]	 2. B is THE **MIDDLE POINT OF** [AC] 3. Draw a circle of **CENTER B** passing through A and 4. …

Fig. 4.9 Asymmetry of the directions of conversions when the language is one of the mobilized registers

We find in this example the basic workings mathematical thinking, the same that Frege had already put into evidence against the epistemological dilemma in which Kant had confined mathematical knowledge: *different designations possible of the same object is the necessary condition to develop a reasoning both creative and rigourous, because it allows a discursive progress "salva veritate"*. Here is one of the three guiding principles of mathematical rationality.[3] It commands all use of natural language, even in primary school, when, for example, students are asked to write messages in order to make construct a figure. Surprisingly, the fundamental operation that consists of producing two different names for the same object is underrated in both the didactic research and classroom. Students are asked to produce verbal explanations and analyze their productions as if using two different designations of the same object was a familiar operation in the spontaneous practice of language. Try to speak following this requirement in talks and discussions with others and you shall see their reactions.

Variations of congruence and non-congruence show that *there is no isomorphism between the representations of a mathematical object in a register and its possible representations in other registers.* It is why the conversions of representations are cognitively non-reversible operations. So, we can summarize the analysis of the cognitive functioning of mathematical thinking, in the following four points:

- There are specific semiotic operations for each register of representation. The creative power of mathematical thinking stems from the variety of the registers used.
- The passage from one register to another requires that we begin to develop a synergistic coordination between at least two registers. This development requires specific activities and tasks, different from those privileged for the acquisition of "concepts."
- The understanding of the "mathematical concepts", unlike the understanding of the concepts in other disciplines involves the synergistic coordination of at least two registers of representation.

[3] The two other guiding principles are the deductive use of theorems and definitions according to the mechanism of the *modus ponens* and the combined use of negation and quantification. The principle of non-contradiction is not a guiding principle, but a requirement of acceptability. To know that you do not have to contradict does not allow to know what and how to say it.

- The non-congruence phenomena outnumber the congruence phenomena. There is no rule to consider them in general and therefore anticipate them, but should be studied case by case, for each activity or every problem proposed.

4.2.2 The Particular Role of Natural Language in the Cognitive Functioning Subjacent to the Mathematical Reasoning

The analysis of the cognitive functioning proper to each register can only be done in reference to a second register. Clearly, this implies that the cognitive distance between the two registers is not too large. In other words, it is necessary that *the variations in the content of the representations of a register can be systematically compared with the variations in the content of the corresponding representation of another register.*

This was the case for graphs and symbolic expressions of relations (equations or inequations), as we just saw. However, the situation changes completely when one of the two registers is no longer a purely mathematical register, and more particularly, when the natural language is the starting register in a mono-representation situation. Natural language is one of the registers used in mathematics to formulate definitions, theorems, mathematical reasoning and to justify solutions. In the teaching of mathematics, the natural language intervenes in all problem statements given to the students, but mainly in problems that require knowledge application. Here is where the first real misunderstanding happen and prove to be recurrent. Students suffer the disconcerting experience that the use of language in mathematics has not much to do with the spontaneous use of the language practiced outside mathematics. This has raised the issue about the best representations to help students understand the word problems. We cannot progress on this issue without asking how the internal hierarchical organization of any discursive production in natural language (Fig. 4.6) can be coordinated with the more simple organization of the representations in the other registers.

There is a considerable cognitive distance between the natural language and other cognitive registers and even the other discursive registers specific to mathematical or logical treatments. This fact makes it difficult to convert the statements of the natural language into representations of the other registers. So, to highlight the variations of statements in natural language (and also the differences between the different kinds of reasoning developed in natural language) with the variation of representation contents, which map one-to-one in the other registers, we need to use auxiliary transition representations. The analysis of the cognitive functioning of natural language as used in mathematics requires to make use transitional auxiliary representations. Of course, not any representation can fulfill this function and it not the same representations that allow us to analyze the problems of application of mathematical knowledge to real situations, and the functioning of mathematical reasoning or even the way of arguing outsize mathematics. Nevertheless, *all the auxiliary representations must comply with the same semiotic condition: to be two-dimensional representations.*

4.2.3 *The Understanding of the Problem Statements and the Need for Transitional Auxiliary Representations*

Let us consider the most elementary example: the statements of additive problems with one operation. What kind of representation can help to understand the following problem that describes a quantitative "transformation" in which the unknown quantity is the one of "the initial state"? Let us look at the two statements, respectively presented to young students and future primary school teachers (Fig. 4.10).

For cognitive analysis of these problems, it is necessary to establish the field of variations of the statements, that is, the field of all the possible statements, which involves some transformation in the way of writing the corresponding numerical equation with a blank for the unknown quantity.

Applying the method of cognitive analysis to the first version of the problem, we see that two discursive variations allow transforming this statement into other problem statements. The first is the place of the missing data.[4] The second *is the choice*

Bruno plays two matches of marbles. He plays *the first* match. In *the second match,* he **loses** 3 marbles. After these two games, he **gains** 6 marbles overall. What happened *in the first match*?		Monday *afternoon*, Pierre **received** 750 F from his parents. He made several purchases, and *at night*, he found he **had** 378 F **less than** *in the morning.* How much did he spend?		
Percent of right answers		Answers given *and maintained after lengthy discussions* by 21-23 years old students in training to teach in the primary.		
			Group I (58 E.)	Group II (24 E.)
8-9 years old	**25%**	Without calculation 378 F	62 %	92%
9-10 years old	**28%**	750 – 378 = 372 F	19%	**8%**
10-11 years old	**47%**	**750 + 378 = 1128 F**	**16%**	
G. VERGNAUD & C. DURAND (1976)		H-P. DELÈGUE & J. ROUSSEL (2000)		

Fig. 4.10 Where does the difficulty of these additive problems come from?

[4]The problems given when teaching are characterized by the search of the missing datum or data. The missing data are the complement of the given data in the framework of the mathematical treatment to be done in order to find it. There are as many types of problems as missing data possible in the framework of each mathematical treatment. It is because these problems are not real problem of research and their resolution appears as particular. Duval, R. (2003). Décrire, visualiser ou raisonner: quels "apprentissages premiers" de l'activité mathématique?. *Annales de Didactique et de Sciences Cognitives*, 8, p. 13–62.

STARTING REGISTER: systematic variation of the STATEMENT		ARRIVAL REGISTER: NUMERICAL EQUATIONS with blank
I. PLACE OF THE MISSING DATA	II. Terms of a semantic opposition marking the DIRECTION of the transformation	Conversion of the semantic term into an relative integer
the final stage (transformation)	win, win ...? lose, lose ...? win, lose ...? lose, win ...?	(+3) AND (+2) = ... (–3) AND (–2) = ... (+3) AND (–2) = ... (–3) AND (+2) = ...
The intermediate stage (transformation)	win ...? win lose ...? lose win ...? lose lose ...? win	(+3) AND ... = (+5) (–3) AND ... = (–5) (+3) AND ... = (–5) (–3) AND ... = (+5)
The initial state	? ... win, win ? ... lose, lose	... AND (+3) = (+5) ... AND (–3) = (–5)
	? ... win, lose **? ... lose, win.**	**... AND (+3) = (–5)** **... AND (–3) = (+5)**

Fig. 4.11 Table of variation of the statements of additive problems

of one of the two terms of any semantic opposition (here: win/lose, win/win). Therefore, we obtain 12 different possible statements for additive problems with a single operation, i.e., *twelve statements that can be converted into a numerical equation with blanks* (Fig. 4.11). There is a third variation that is always considered the most important didactic variable: the size or magnitude of the numbers associated with one of the two terms of the semantic opposition (column III). In fact, it has no role in the analysis and understanding of the problems. Therefore, it is necessary to neutralize it.

Thus, to solve an additive problem *is to convert the statement into a numerical equation with blanks.* The difficult point in this conversion is that it is necessary to choose the operation marked by the operator "AND" in the numerical equation with blank: addition or subtraction? The term of semantic opposition "win/lose" (Fig. 4.11, column II) only designate the fact that the numbers are positive or negative integers.

The analysis of the students and future teachers' answers shows that the two terms of semantic opposition mean only an addition and subtraction, respectively. So, the choice of the arithmetical operation for the operator "AND" is always done according this common meaning. The answers are divided into successes and failures, according to the congruence or non-congruence between the statement and the arithmetical operation to be chosen. The congruent interpretation leads to a right answer when the question refers to the final or the intermediate stage. However, this is not the case when referring to the initial stage when numeric data are associated with the two antonyms (the last two lines of the table). The most interesting fact is not the high failure rate, but inability to understand the solution when it is then explained. In both groups of future teachers (Fig. 4.10), the mathematical response was systematically rejected when suggested in the final phase of the session. They were convinced that the mathematical answer was not possible. Neither the teacher nor the few students that have understood were able to make change them after a discussion that lasted more than 30 min! And, these future teachers had just spent 3 years in the University.

The cognitive distance between the statements of additive problems and the numerical equation to be recognized to perform the calculation is, therefore, considerable. It is necessary to resort to an auxiliary representation to understand the resolution of all additive problems, mainly those with non-congruent statements. This should allow to separate the statements in the two types of information relevant in the verbal descripton of the dataset: this about the succession of the stages and this about the term associated with a numeric value (respectively, in italics and bold letters in the two statements of Fig. 4.10).

R. Damm built a graphical representation that meets this requirement.[5] The two types of information are respectively located in two semantic axes, and a semi-line parallel to the axis of numerical values materializes each phase. In other words, we have as many parallel numerical axes as many stages described in the statement of the problem (here three). The information relevant to solve the problem appears as the successive intersection points determined by each kind of information. Then, we can plot them on this chart as we are reading each sentence of the problem statement. The procedure is as follows:

- We mark the numerical value on this first parallel line, and then we transfer it on the following parallel line that materializes the second stage (dotted arrows).
- We start from the position marked on the second parallel line to mark the second numerical value (up or down vertical arrows).
- When we reach the parallel line corresponding to the missing numerical value, we extend the transfer to the next parallel line
- The last transfer gives the answer if the missing numerical value corresponds to the final stage. Otherwise we have to transfer back the last position obtained to the parallel line we have jumped. The distance upwards or downwards between the two marked positions allows to read the result.

[5] Damm, R. (1992). *Apprentissage des problèmes additifs et compréhension de texte*. Thèse ULP. Strasbourg: IREM.

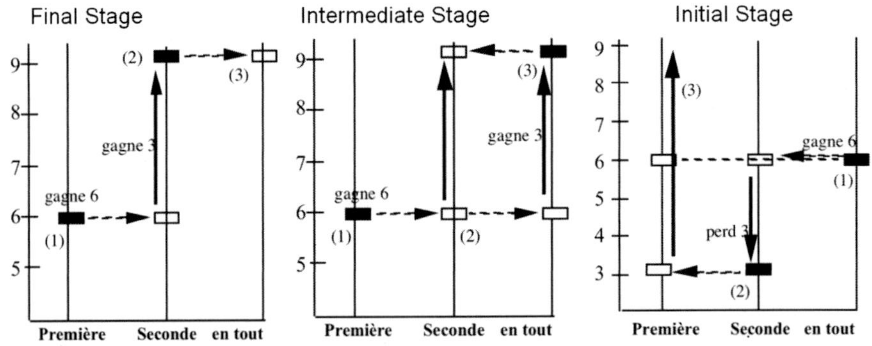

Fig. 4.12 Variations of the graphs according to the place of the missing numerical value

See an illustration of this procedure. We obtain three different graphics depending on the location of the missing data.[6] In the diagrams below, the black rectangles correspond to the marked position and the white rectangles correspond to the transferred positions (Fig. 4.12).

In his work with young students, R. Damm designed buildings in place of the three parallel lines, in order to represent the successive going up and down the floors of a building. Each displacement can then be physically conducted using a magnetic board and various coins. A student reads the first sentence of the statement, and the other should put the coin in the corresponding position. The activity can be reversed. A student puts the coin, and the other must describe the location or move of the coin.

Students can then learn *to recognize the units of meaning in the statement, which are relevant to select the arithmetic operation to be performed.* By varying themselves their formulations, they can separate the two kinds of verbal information, realize how take into account the place of the unknown quantity and observe on this bi-dimensional representation how all these aspects of the problem coordinate with each other. They understand quickly almost all statements of the additive problems and the two incomprehensible problems of transformations (above Fig. 4.10) become problems that are not more difficult to understand than all others. Moreover, without any teaching about the relative integers. Such auxiliary representation is of course a transitional representation. The students abandon it as soon as they understand because its use seems to them a slow and costly procedure.[7] But, this cannot be decided by any didactic organization of activities.

We can make the students understand and solve all the additive problems well before teaching about the relative integers. Moreover, that this understanding is one

[6] Duval, R. (2005b). Linguaggio, simboli, immagini, schemi... In quale modo intervengono nella comprensione in matematica e altrove?. *Bolletino dei Docenti di Matematica*, 50, p. 19–39.

[7] These explanations and points made by the students led us to talk about representations of “transitional” in the sense that Winnicott had spoken of “transitional objects” a reference to the objects that young children need to have always in situations where they feel lost.

of the prerequisite conditions for a first discovery of algebra. It allows introducing numerical equations with blanks, which are the simplest algebraic equations, where the "…" blanks to fill in, are the equivalent of an unknown letter (Fig. 4.11, above). Their resolution requires that we perform one of the elementary and fundamental operations of algebra: to pass a term from one member to another by changing sign.

4.2.4 The Problem of Cognitive Connection Between the Natural Language and Other Registers

In the statement of a problem, the distance between the common and the mathematical use of the language is not due to the use of a particular vocabulary. But, to the specific way of verbally describing the problem data: one needs to cross two different semantic determinations because *the conversion of the statements into register of symbolic writing cannot be straightforward*. It is often necessary to use an auxiliary transitional representation to allow separating two semantic dimensions. We have just seen in the case of additives problems, and the same thing is true for all the word problems. Evidently, not the same representations are used when dealing with multiplicative problems, putting data into equations or, more generally, applying some particular mathematical knowledge to a real situation. In all these problems the crucial point in the use of natural language is about the object designation operations. They are quite different in mathematics than outside mathematics.

The distance between the mathematical and common use of the language appears deeper and ever more difficult to overcome when used to enunciate definitions, theorems or deduce, because it is necessary to distinguish between a theorem and its reciprocal, or between a valid reasoning and a rebuttable argument, recognize a vicious circle, etc. That depends on discursive operations, which are specific to mathematics. The trouble is the use of the same linguistic markers, "logical" connectors and sentences with "that" clauses (called "propositional or epistemic attitudes") in mathematics and outside mathematics, whereas the underlying discursive operation are quite different for defining, arguing and even applying principles and rules in jurisprudence. We have just to look at negation to realize that. In mathematics, its use is always articulated with quantification to produce statements that are as informative as the opposite statement, which appears denied. The use of a formal languages allows to make this articulation more operative and introduce the double quantification. On the contrary, in spontaneous language use, the functioning of negation and quantification are varied and independent of one another. Moreover, the description of concrete situations, stories or "reports" are made without resorting to negation, because its use signifies refusal, introduces a rebuttal, or expresses a lack of something.

These distances are because they are not the same discursive operations that we mobilized and developed in spontaneous speech and the mathematical discourse. And, it may be because everything has been done to reduce the place and role of the natural language in the practice of mathematics. The development of mathematical logic and formal languages, following the pioneering work of Leibniz, Boole, Frege, Russell, witnesses the search for a *semiosis* that would remove the natural language in favor of formal language. It is, then, a *semiosis* that only contains a single representation register, excluding not only the natural language, but the geometric visualization, and ignoring the graphic representations. One of the phantasms of the reform of modern mathematics was the attempt to replace, the complexity of natural language (and also elementary geometry) with the propositional calculus, first-order predicate calculus and linear algebra. The other was to make the students discorver all the notions and concepts to be learned, by following their mathematical order of construction. So the priority was given to the logic of sets and relations to prepare the notion of function since the beginning of the middle school (11 to 14-year-old students). There was, certainly, the use of some auxiliary representations, Venn diagrams, diagrams, arrows. But, they were just illustrations of sets operations or one-to-one relations between two sets, although their use required a mathematical comment in natural language!!! An inconsistency that was amazing from both educational and cognitive point of view. And, at the same time, linguists used the predicate calculus or formal languages (monofunctional discursive register) to analyze the production of phrases in natural language (multifunctional discursive register).

Between these two types of discursive registers, natural language and symbolic expressions of relations, we saw a considerable cognitive distance and the need of auxiliary two-dimensional representations to jump it. More generally, we saw that the analysis of the cognitive functioning that underlies all discursive productions could only achieved by converting first the discursive production into a non-discursive register, even for the understanding of all valid mathematical reasoning.[8]

4.3 Functional Variations of Phenomenological Production Methods and Semiotic Representation Registers

Mathematics require to be written, even when we are "telling" them, i. e when they are orally explained. Therefore, a blackboard is necessary for class. In a restaurant, we scribble an outline or a calculation on one corner of the paper towel. On the contrary, the poetry has to be spoken in a low or loud voice and listened to even when we read it, alone for ourselves. It has a rhythm that needs to be understood and memorizes the words as in a song. Physics require to use instruments for measuring time, angles or distances to observe the movements. It also requires to record the measurements obtained and make again new observations. Thinking as "mental activity" seem go against all these kind of activities. *It is always silent as so far it is raised by an issue that cannot be reduced to any formulation or any possible observation.* In a debate, in contrast, the fact of not being able to reply immediately to an objection, is perceived as

[8] Duval, R. (1995). *Sémiosis et Pensée Humaine*. Berne: Peter Lang.

a lack of argument, and this invalidates *ipso facto* what has just been said. It is leaving the "last word" to the opposite point of view.

Each one can continue this enumeration of the multiple forms of the human cognitive activity, according to the various areas of knowledge and moments of working mode proper to each area. All involve the production of representations, which are semiotic or non-semiotic. The analysis of the observable or experienced forms of the human cognitive activity raises *the question of the relations between the semiotic and non-semiotic representations, within the overall cognitive activity of each individual.* It raises at the same time the question about the nature of the non-semiotic representations. To answer these two questions, the cognitive and educational theories are confined to re-actualize the classical philosophical opposition between *concepts and perceived reality* into the oppositions between *mental and material* representations, or *internal and external* representations. In this very general description of the cognitive activity of individuals, the semiotic representations are then regarded as external representations and opposed to mental representations that would be non-semiotic. There is thus two incompatible cognitive approaches, at least with respect to mathematics and all the comprehension problems that its teaching raises. The approach that is commonly shared relativizes the importance of semiotic representations subordinating them to mental representations never directly accessible, while the other gives semiotic registers a central role in the development and exercise of the mathematical thinking.

4.3.1 The Misleading Confusion Between Functional and Structural Variations in Production of Representations

The difficulty and complexity of every analysis of global cognitive activity of individuals are due to the fact that one must always distinguish two radically different aspects in the productions: the phenomenological production modes of representations and the different systems mobilized to produce them. The methodological necessity of this separation becomes crucial when the analysis of the cognitive activity concerns the acquisition of scientific knowledge, because it is necessary to take into account the mode of access to scientific knowledge objects. So we see that *there are several possible phenomenological modes of production of the semiotic registers.* This is obvious for the natural language: it allows three phenomenological modes of verbal production, as Vygotsky (1933) had stressed in his work. Similarly, the mental calculation mobilizes the decimal system, i.e., a semiotic system, and a non-vocalized verbal evocation of numbers to keep them in mind while we are doing calculation (Fig. 4.13).

This table below does not include the diversity of scientific instruments that allowed to increase the power of visual discrimination, from the seventeenth century and, therefore, broaden the field of perception, because none of these instru-

		PHENOMENOLOGICAL PRODUCTION MODE			
		MENTAL	ORAL	GRAPHICS/ VISUAL	COMPUTER MONITOR
KIND OF SYS-TEMS OF REPRE SENTA-TION PRODUC TION	SEMIOTIC REGISTERS INTENTIONAL PRODUCTION: *reference relationship*	**NON-VOCAL VERBALI-ZATION** *Functions of objectification (to oneself) and fast treatment, but reduced*	**SPEECH** *Functions of dialogic /rhetoric communi-cation, or objectivation*	**WRITING, DRAWING** *Functions of non-restricted treatment, but slow, objectification and communication*	**DISPLAY BY USING A PULL-DOWN MENU** *Function of Instantaneous and umlimited treatment, or simulation*
	NON SEMIOTIC Neural networks sensory receptors AUTOMATIC PRODUCTION: *causal relationship*	**MENTAL IMAGE** (memory) *Function of immediate objectification*		**REFLEC-TIONS, PHOTOS** Functions of materials or recorded data	

Fig. 4.13 Functional and structural variations of the representation producing systems

ments can be used to gain access to mathematical objects. This table allows several important observations.

- *Semiotic representations are neither mental nor material, neither internal nor external.* They are only semiotic. There is an absurd way of opposing mental and semiotic representations according to the traditional philosophical opposition that would led to believe that a calculation would not be semiotic when it is done mentally, but becomes semiotic when it is written down!
- *The nature of a representation depends on the system that produces it.* In other words, the diversity of all representations involved in the global cognitive activity of individuals is done based on the systems that allow producing them (Chap. 2). The general cognitive principle that assumes the need of a specific system of representation production per cognitive function distinguishable does not apply to the registers. Not only the natural language but also the one of configurations of 2D dimensional shapes can fulfill quite different cognitive functions (Fig. 4.6, above). Thus there are several registers possible for the only function of calculation, i.e. algorithmic substitution of semiotic representations with each other.
- Being interested first in the phenomenological modes of production in the analysis of cognitive activity is *to prioritize a functional analysis of the production of representations*, i.e. to focus on the reason what they are produced or what an individual meant by producing them, whatever the mode and register mobilized. On the contrary, analyzing the registers mobilized in the representation production is *to prioritize a structural analysis*, i.e. to focus on the meaning units and intrinsic transformations, which are specific to each register mobilized in the production of representation. The structural analysis of the students' production is more signifi-

cant and conclusive than functional analysis to diagnose their recurrent difficulties in understanding and work out the crucial factors to make each student learn to learn for oneself mathematics.

- Three points of view should be considered in the mobilization of a semiotic register whose production is intentional, that is, consciously conducted and controlled. From a mathematical point of view, the mobilization of a register depends on the explicitation and treatment possibilities it offers. From a psychological point of view, it also depends on the quickness and power of production that each phenomenological mode allows. The number of meaning units that can be kept together immediately and overtime characterizes them. *It is there the fundamental difference between on one side mental or oral production, and on the other written production.* The written productions open up a field and development possibilities for the cognitive activity that is much wider and more powerful than those the only oral and mental productions do. So there is a purely oral math that are quickly faced with the impossibility of development, and there is the mathematics that have being developed first with writing, then with the invention of new registers of representation. At last from a cognitive point of view, the mobilization of a semiotic register depends on the level of coordination of registers, i.e. the level of awareness of the meaning units change according each register and transformations intrinsic to each register.
- *Oral, i.e., the vocal output or speech can fulfill two functions.* Its essential function is, of course, the dialogic COMMUNICATION, one in which the interlocutors are alternatively listeners and speakers. This alternative position is weakened or suppressed when the speaker addresses an audience. It is then more a rhetorical communication than a dialogic communication. The second major function of speech does not depend on any communication situation. An immediate communication situation can even parasitize it. This function is OBJECTIFICATION. In this case, one expresses oneself solely for oneself. Then the expression raises an awareness of what one was not before the expression act, or allows distancing oneself from an emotionally very violent experience. The requirements for the content and accuracy of expression are not at all the same when the act of speaking fulfills either a communication or an objectification function. But depending on the circumstances, the speech fulfills more or less each of these two functions! Hence an ambiguity in classroom when, for example, the teacher reacts to the mathematical inaccuracy of what students say, as if the spontaneous speech of each student was not the first cognitive step necessary to become aware of what the teacher aims at making understand.

4.3.2 The Computer Monitors: Another Phenomenological Mode of Production of Representations

From the "symbolic" machine of Turing (1936, 1937) to the building of the first computer (1943, 1946), and after the microcomputer has appeared on the market in the mid-1970s, there was a tsunami in all activity sectors of society from the 1990s until today. In the teaching of mathematics, the monitor and keyboard came to replace the paper and pen, and also … the blackboard. What are the contributions and, eventually, the limits of its use in mathematics education for developing the global cognitive activity of students?

Paradoxically, there is little actual research on this issue, considering that the computer is a powerful tool, which is both fun and simple to use. In fact, we are still in the enthusiastic phase of innovation and, as in any real innovation, we hope for pedagogical or didactic miracles. We will present here only some elements to target the key points for research on this issue.

First, it is important to place precisely the software contribution in comparison with the three phenomenological modes of semiotic representations, *which are entirely dependent on the individual cognitive development.*

From this perspective, we can make the following three observations:

- Computers do not constitute a new representation register. For one simple reason: *the representations they display are the same as those produced graphically on paper for a visual apprehension.* Seeing any geometric configuration on the screen requires the same dimensional deconstruction and visual discrimination of the merological transformations possible than seeing it on the paper. Likewise, seeing graphs on a monitor requires us *to be able to recognize the mathematically relevant visual value*s *and coordinate them with the meaning units of the corresponding equation or inequation.*
- However, *they constitute a radically new phenomenological mode of production, based on the powerful acceleration of treatments.* They display on the monitor almost as quick as a mental production, but with unlimited treatment power compared with the possibilities of the graphic-visual mode. We obtain immediately, much more than anything we could get by writing calculations or constructing geometrical figures and graphs, for several hours or days.
- The most spectacular phenomenological novelty is that *non-discursive semiotic representations* become manipulable as real objects. We can move them, make them rotate or extend them from one point. This "dynamic" aspect is just a consequence of the computer capacity of treatment, which is considerable in comparison with the other phenomenological modes of production. *That allows to meet a new epistemological function that the others modes of production cannot do: exploration by simulation.* Extremely important in sciences, this simulation function allows the heuristic exploration of mathematical problems and even proof as for the theorem of four colors. But, this requires a kind of proof based on a method that uses only algorithms.

The important question for all that concerns the basic education is the degree of cognitive activity that the use of a computer actually requests from a non-expert person. In other words, what does a student just need to do to display on the screen what can be a visual response to a question? We can then see that the real interface between the computer and the individual is not what it displays on the monitor, *but what allows commanding display, that is, the pull-down menu for the instructions.*

PULL-DOWN MENU	ACTION	MOBILIZED COGNITIVE ACTIVITY
• A list of few terms designating some properties, operations or mathematical objects.	• Choose a term for an instruction or compose a sequence of several instructions.	• Knowledge of mathematical terms and *decomposition of the expected figure into several instructions corresponding to the terms of the list*
• Empty space for writing digits or symbols.	• Write an equation.	• *Preliminary coordination to choose the kind of equation that allows obtaining the expected curve type or surface.*
• Rows of icons.	• Support over an icon.	• *Recognizing the icon that encodes the instruction* corresponding to the expected display.
• The mouse or the track pad, or the screen.	• Move with the hand or a digit.	• Automatic pilot for coordinating move and instruction's localization on the screen monitor

Fig. 4.14 Analysis of the cognitive tasks required to use a computer

We can then analyze the cognitive tasks required for the use of each software depending on the actions that the menu allows or excludes (Fig. 4.14).

The pull-down menu acts as a very restricting, or even as a reducing filter, which goes against the way of seeing or wording in the three other phenomenological modes of production. And that leads us to make four important observations:

- Even if we can multiply instructions effortlessly, we introduce a linear sequential visualization in the monitor similar *to the sequence of sentences* in speech. This means that we found limitations of immediate memory that are proper to the distracted or attentive listening. It is, therefore, difficult to perform an observation or comparison of variations of graphical representations with respect to changes in the writing forms of the corresponding equations. *This work of observation and comparison requires a synoptic view of both kinds of change made.* In other words, this work requires that their respective co-display remain as a whole in front of the students in order they can focus back and forth between the two kinds of variations. This can be realized easily in a graphical-visual way, and confining oneself to some elementary representations.
- A pull-down menu singles out a representation register to obtain the corresponding representation in another register. For example, it singles out the entry by writing the equations. But the entry by the graphical representations to obtain the corresponding kind of equations would require other software. However, both entries are also needed to develop the cognitive coordination between these two registers, in the learning phase.
- To construct a figure, the instructions are terms of geometrical objects or properties. That requires anticipating the deconstruction of the 2D configuration to be constructed into 0D and 1D figural units and also 2D subconfigurations. *This decomposition, which is imposed by the few terms of the menu, can be strongly non-congruent with the way the eye recognize the possible decompositions of the 2D configuration.* Furthermore, the verbal menu introduces a restriction that

does not originate from the visualization: the need of naming the extremities of the 1D figural units selected in the lexical menu.

- The interface with a computer eliminates language, i.e., all discursive operations. *Its use reduces the use of a lexicon or keywords.* We can even, short-circuit them completely by replacing the words with icons. In fact, *the display mode on the monitor is only innovative for the registers other than natural language.* The "word processors" only help in formatting, that is, the editing work and not in the discursive production. The development of the personal discursive production is as crucial in mathematics as in other knowledge areas, at least for the period of basic education.

4.4 Method of Analysis of the Activities in Class and Student Productions: The Problem of Didactically Relevant Variables

The representation registers allow defining the cognitive factors that drive the understanding and misunderstanding during the learning process of mathematics. Nevertheless, are these cognitive factors didactic variables? They can be expressed in terms of organization of recognition tasks when exploring experimentally the cognitive processes of mathematical thinking, but are they useful for the teacher when organizing activity sequences in class to make students acquire mathematical knowledge? Many refuse to consider the registers as didactic variables and think that the only important factors are those related to mathematical concepts and research of situations, tools, activities that can really motivate and prepare the learning of these concepts. However, for most students, such a way of teaching results in the lack of capacity to use the learned concepts outside the particular context in which it has been taught. This lack of subsequent availability of the mathematical knowledge learned raises the real question for mathematics education: what do we want the students to achieve cognitively, and not only mathematically, when we organize a "didactic sequence"?

4.4.1 The Organization of Sequences of Activities Always Has Two Sides

One side of mathematical activity is evidently the mathematical property that should be discovered and acquired by students. Several kinds of tasks to be performed are chosen so that they cannot be carried out without implicitly using this mathematical property. The teaching sequences are constructed according to the general scheme of development in which learning starts from practical or specific activities and progresses towards theoretical or formal constructions through quite different verbal interactions.

The other side is that of the representation of registers and phenomenological modes of production that will be favoured during the sequence. And the key issue of the interpretation is then *the internal transfer from one phenomenological mode to another and from one register to another.* Here things get complicated. It is commonly assumed that these transfers do not arise deep difficulties or should become spontaneous with the acquisition of mathematical concepts. Is this assumption well founded in mathematics education? When we are in a class and observe regularly over a period of 6 months, not what the teacher does or says, but what some students do, it is not only the basis of observations that is modified, but also the criteria for analysis and interpretation concerning the success or failure of learning. We are then faced with the following questions:

- *(Q.1) Are all students or only some of them successful in each of the activities of the sequence?*
- *(Q.1′) What happens to those that are not successful? Is the pooling of students' productions by the teacher with the whole class enough so that these students can perform the next activity?*
- *(Q.2) What is really common between two successive activities that allows transferring to the second what has just been successful in the first? This cannot be just the mathematical property since acquiring it is the goal of the sequence.*
- *(Q.2′) Are the students aware of what allows changing from one activity mode to another or DO THE ACTIVITIES REMAIN FOR THEM SEPARATE TASKS?*

These last two questions are the most important for determining the didactical variables. They highlight the need of a cognitive analysis of each task of a teaching sequence, and not only mathematical analysis in terms of problem-solving procedure. This leads us to reformulating the (Q.2) and (Q.2′) questions as follows:

- Is it good enough to do *locally* all the tasks of any didactical sequence to understand and acquire knowledge taught?
- Does the mathematical activity explicitly required *create a resonance effect on the cognitve activity implicitly required?* This question concerns the comprehension in terms of coordination of registers, i.e. recognition of what is relevant, and not mathematical justification.

Registers are the tool to analyze whether a teaching sequence meet the necessary conditions to develop the ability to work in a mathematical way. They allow defining what is cognitively required so that the students can achieve an understanding of the mathematical knowledge to be acquired. They allow also assessing their productions and diagnosing their blocking points.

To illustrate these points, see an example of analysis. It concerns a sequence of activity that has been proposed to 10 and 11-year-old students for the acquisition of axial symmetry.

4.4.2 The Field of Work Cognitively Required for a Teaching Sequence of Geometrical Activities at Primary School

The cognitive field of a teaching sequence in geometry has two dimensions.

The first concerns the aspect of geometric objects that is favored for the activity to be achieved: the aspect *shape* of a configuration and its specific operations of recognition and dimensional deconstruction, or the aspect *magnitude* and its operations of measurement and calculation. Every activity that mixes up these two aspects creates *ipso facto* the impossibility of entering the unnatural way of seeing that is required in geometry.

The second dimension concerns the temporal unfolding of the heterogeneous tasks of the teaching sequence. It contains (I) the manipulation tasks of (2D/3D) physical objects in relation to the mathematical property, then (II) working out on the 2D/2D figural representations, with emphasis on the procedures based on axial symmetry either for constructing a figure or verify that the figure has this property, and finally (III) wording geometrically what the students are supposed to have developed during the preceding tasks.

The margins of the table below translate these two dimensions (Fig. 4.15). Each cell inside corresponds to the relevant operation favored by each particular task. From (I) to (II) there is a jump that is represented by an empty column, and from (II) to (III) there is a change of register.

This table represents a teaching sequence that spans several months, once per week, and that has been proposed in a class of 10 to 11-year-old students. It allows analyzing *a priori* whether it meets the necessary cognitive conditions to develop comprehension in geometry.

- We find the general didactic scheme of development in which learning starts from practical or specific tasks (stage I and II). They aim at focusing the students' work on the recognition of the relevant figural units. They have all being designed to make them realize the dimensional deconstruction of the perceptually dominant 2D–shapes: solving the posed problems used operations requires analysing 2D figures in terms of 1D and 0D figural units, that is, intersection points obtained by prolonging or aligning points (Chap. 3, Sect. 3.2.2). Every operation on the magnitude aspect is excluded.
- *There is nothing in common between the tasks of the stage I and those of stage II.* The organization of this teaching sequence was based on the classical expectation of continuity and transfer from one stage to another. It was assumed that the variation of the tools should allow students to enter more easily in the activities of the stage II. It was above all assumed that both stages I and II should, by themselves, *allow an understanding without language.* The language would come into play only later (stage III) to put into geometrical words the property, which students have already been applying and should have understood.
- Their no real place for the language in this process of conceptual learning. Its role is at the end of the teaching sequence and, *above all, reduced to the operations of*

	I. **MANIPULATE** the 2D/3D PHYSICAL OBJECTS		II. SEE the 1D and 0D figural units in the configurations of 2D shapes		III. VERBAL DESIGNATION of the relevant figural units and its relationships	
	material ACTION	**FOLD** *to overlap*		Use 1D OR 2D **INSTRUMENTS** FOR 1D DRAWINGS	ORAL MODE of production	WRITING MODE
SHAPES	*reversible OPERATION*	**REVERSE** *the transparency support to* **OVERLAP** *it on the starting figure*	?	• **PROLONG** UP to an intersection point • **ALIGN THE POINTS** • **OPPOSITE ORIENTATIONS** of the 1D units from an axis • **TRANSFER OF** A length without graduated ruler	REDUCED DESIGNATION OF THE STROKES to be done (or realized) on the figure	Wording of AN INSTRUCTION message *to make the figure be constructed by other students.*
MAGNITUDES	No	No		No		No calculation

Fig. 4.15 The cognitive field of a sequence of activities about axial symmetry

designation of figural units or relations between figural units. The stage III aims at introducing the basic geometrical lexicon. These operations of qualifying designation are first requested only in an oral mode. The writing mode stems from the teacher's initiative when he writes on the blackboard what the students have to memorize and copy into their notebooks so that they can reuse it in further written exercises or problems to solve. The tasks regarding the elaboration of instruction messages were requested only exceptionally (Chap. 3, Sect. 3.2.1).

4.4.3 Observation of the Students and Analysis of Their Productions and Reactions

I regularly observed for 6 months the unfolding of the sequences of activities in geometry, including those about axial symmetry. The class was divided into two groups for this teaching. And each group attended the same teaching delivered by the same teacher in two consecutive sessions. Thus, I could observe more closely the work of two students in each group and also check out from time to time what the other students were doing. I observed what was said, done or not, as well as exchanges between the teacher and these students or facing the group. Several observations were then stood out. And, I could check and corroborate them when the same teacher used the same teaching sequences with other students in the following year.[9]

[9] Duval, R. (2014). Ruptures et oublis entre manipuler, voir, dire (oral) et rédiger des instructions. Histoire d'une séquence d'activités. In C. Finck Brandt and M. Thadeu Moretti (Ed.) *As Contribuições da Teoria das Representações Semióticas para o Ensino e Pesquisa na Educação Matemática*, 227–251. Ijui: Unijui.

There was no transfer from material manipulations (I) to the instrumented constructions of Figs. (II). The shapes to reproduce or to complete (pine tree, two triangles superimposed in the shape of a butterfly, etc.) have been elaborated to verify whether they were symmetrical or not, then find out their axis of symmetry, at last use the property to construct symmetrical configurations. *This lack of transfer from material manipulations to the construction of figures is not at all surprising, because there is no relationship between operations made in 3D and those required in 2D, nor between their criteria of symmetry recognition.* Material manipulations could not help students neither to see the strokes that needed to be drawn, nor recognize the transformation underlying the apparently unchanged positions of opposite points in symmetrical shapes.

But above all, there was no transfer from material manipulations or instrumented construction to the understanding of oral explanations by the teacher either to formulate the instructions for constructing or recognition criterion of the property (Stage III). In both cases, the incomprehension concerns the use of the geometrical terms to designate the 1D and 0D figural units (segment and points).

When we analyzed what the students tell what they had done individually during the class work, *the most striking point is not not what they explained but what they don't mention at all because they had not noticed it.* The crucial point in reversing the transparency (Fig. 4.15, col. 2) was not the coincidence of the shapes (starting figure and its reproduction on the transparent) but the exchange of the positions of the vertices resulting from this reversal. *No student mentioned it, even if they had done spontaneously and mentioned* this important use of the transparency. The teacher had to remind the students this central operation to raise the awareness of symmetry.

Here we touch on one of the essential conditions for cognitive learning. *The awareness of the operation involved in a concrete action can only happen when we express both the operation done and its result.* Thus, the result of the reversal to superimpose lies in the fact that the branches of the pine tree that were on the right of the starting figure are on the left in the reversed figure, and vice versa. In the proposed activities, the result that drew attention was that of the construction task: to complete the missing branches in the pine tree. The overlapping intervenes only in the research of the coincidence of the contours, regardless of the rotation in space or the translation in the plan. Therefore, *the students could not become aware of the link between this operation and obtaining the contour of the "symmetrical half", nor the correspondence between two opposing points on both sides of the axis of symmetry*. The necessary condition for seeing in terms of "a point and its image." had not been taken into account in the teaching sequence (Fig. 4.15, col. 3).

A significant distance appeared between *what the teacher was saying orally* and *what he wrote on the blackboard* for the students to copy in their notebooks and use as reference. On the first year, he introduced the explanation of the object-image relationship orally and wrote down everything that concerned the constructing procedures to complete the configuration by using the axial symmetry. In the second, on the contrary, the geometric vocabulary was introduced in the formulations written at the end of the teaching sequence, while in the oral exchanges during the stage

II the teacher focused especially on the more concrete and more pragmatic vocabulary used by the students.

This significant distance appeared also in what was asked of the students concerning the designation of the various figural units. The level of discursive explicitation is not the same according to the phenomenological mode of production. In the oral exchanges to enter an activity, or during verbal interactions, *the response expected by the teacher is almost always limited* to a single word, *the right geometric term.* The request for a written production was somewhat exceptional. It was limited to the production of messages to "pass the information on to another student" for constructing some configuration. During this activity, which had been requested after several sessions, most students had great difficulty. The shift from speaking to writing a message was also a very big leap for these students. Most students encounter this difficulty. In geometry, *writing a message requires us to carry out three different discursive operations that are not all performed when we speak*: designate the 1D figural units to be drawn; to qualify its relation to another figural unit of the configuration, and associate the successive instruction statements to avoid a gap to the one who will reconstruct the figure (Figs. 3.2, 3.3).

These observations help clarify the questions about the transfer from a kind of activity to another (Q.2) and (Q.2′). They bring us to the crucial educational issue about the relationship between action and verbalization in learning of mathematics. What is the lacking activity (Fig. 4.15, empty column) that could favor the awareness and transfers that are not realized?

4.4.4 Interactions and Cognitive Impact of Three Kinds of Verbalization on Understanding

Specific tasks of verbal expression are needed so that the students can evolve from the description of the concrete actions and usual designation of shapes perceptually recognized to the mathematical formulation of instructions linked to one another. A first task focuses on the *relevant operation involved in the handling action of material support.* A second focuses on each of the two operations of designation that students meet for the first time and will never practice outside mathematics.

The importance of these two intermediate tasks is due to the existence of an implicit or silent verbalization. We can indeed speak of "SILENT VERBALIZATION" as we speak of "silent reading," with no oralization or movement of the lips. This silent verbalization is close to the spontaneous oral expression, and the spontaneous oral expression is just manifesting it as Vygotsky showed. *This silent verbalization is important because it commands the management of intentional activity, as well as understanding of what is obtained when acting.* By this silent verbalization, language does not follow activity, but it comes permanently with it allowing us to anticipate or manage the advance of any activity. Every mathematical wording, or understanding of mathematical utterance, requires a deep change in this spontaneous silent verbalization.

These intermediate tasks of verbal expression should aim to encourage this change because *it can only occur through a work of explicit formulation of the activity with which it is in phase.* In the teaching sequence about axial symmetry, it would have been needed to make students:

- Realize the operation of reversal of the transparency and above all its outcome, which does not consist of the perceptive coincidence the shape outlines, *but the right-left exchange of the 0D units* (the vertices, for example). This task has to be achieved *during the stage I by using the student words.*
- Become aware of the three designation operations proper to the geometric language, *as well as the double designation of the same objects* (Chap. 3, Sect. 3.1.2).This task has to be achieved during the stage II, before introducing the geometrical terms about axial geometry, and the stage III before requesting for writing a message of instructions.

These intermediate tasks should precede the introduction of new terms, as well as writing of mathematical utterances on the board and in the notebook for further learning. They are the cognitive conditions to make the students evolve from their spontaneous silent or oral verbalization to the relevant formulations required for justifying and solving problem in a mathematical way. The understanding emerges when the silent verbalization integrates the different discursive operations that the mathematical discourse merges into technical words, which condense definitions, and theorems for valid reasoning (Fig. 4.6).

Finally, the cognitive field of a teaching sequence presented above can be developed by introducing new columns that concern the language.

4.5 Conclusions

Research on the nature of specific difficulties of understanding with which the learning of mathematics is faced, and the conditions of mathematics education *for all students* require two aspects to be considered. The first is the completely apart epistemological situation of mathematics in the access to the objects it investigates. The second concerns the material based on which we can observe and describe the understanding/misunderstanding phenomena: the productions of the students.

From a methodological point of view, consideration of the students' productions from a research perspective raises two fundamental questions. The first is the collection of data: how to collect the students' productions so that they constitute reliable data? The second is that of interpretation: how to analyze the collected productions, so that their interpretation is objectively controllable and is not just a teacher evaluation as any correction in class or at an exam?

The first question retains more generally the attention. It concerns the choice of the frame of observation: a work session in the classroom, or outside, assessment questionnaires for larger populations. It also concerns the fact that the data are collected by recording in live a work session in a classroom or are a corpus of student's written personal productions over a period of several months. The device of observation can be a teaching sequence or set of micro-tasks prepared by the researcher. A set of micro-tasks can be organized according the two basic experimental requirements: clearly identify the factors and only changing one while keeping the others constant. The independent variables are the cognitive or didactical variables, and the dependent ones are the performances in the micro-tasks proposed. But, regardless of the chosen frame to collect data, the question of their analysis and their interpretation is the methodologically crucial issue. It is at the heart of the problem of the validity and contribution of educational research, which have multiplied exponentially in the last 30 years.[10]

It is essential not to confuse two types of analysis. There is a mathematical analysis of the productions according to the particular knowledge that students should mobilize in the proposed activities and problems. Its acquisition is the goal of education. There is the cognitive analysis of the productions in relation to the representation registers that the resolution of the proposed problems requires mobilizing. It consists of decomposing the solution into the implicit or explicit conversions required to get a representation of the problem data in the register whose intrinsic transformations lead to the solution. This *a priori* decomposition is the analysis tool to highlight the crucial points about the student's comprehension or incomprehension of the way of working in mathematics. The cognitive analysis presents two methodological advantages. It allows to analyze precisely not only everything a student does, says or works out, but also everything that the student does not do, or what he/she does not notice even when he/she does it. Subsequently, it allows *comparing the productions of the same student* in problems that mobilize very different mathematical knowledge. It is only at the level of a comparison overlooking productions with different problems and for longer periods of time that an in-depth interpretation in terms of understanding and knowledge acquisition can be made.

The need for cognitive analysis compared to the mathematical analysis is because what appears to be a success from a mathematical point of view is always a local and circumstantial production. This does not mean further successes in other situations or acquisition which will enable to learn new concepts. Similarly, the mathematical analysis of the errors does not return to its root causes. This is particularly true for the recurrent and systematic errors that most students do not overcome and are almost always "off the field" of most research, because they are not directly associated with the "concepts" taught. Futhermore, taking into account errors is very superficial, because there are blocks and school mathematics dropout, which the students experience almost always as a failure more significant than the mistakes they make. The registers provide a tool to analyze the transformations of representations and knowledge transfer which

[10] Duval R. 2015; Cuestionamientos sobre la "elecciốn" y utilizaciốn de teorías en Mathematics Education. *Didáctica de la Matemática, una mirada international, empírica y teorica* (Bruno D'Amore. Martha Pinilla Ed.) p. 159–182. Universidad La Sabana.

the students perform and, above all, those that they cannot achieve in the various activities and problems throughout their schooling.

The students' productions in mathematics are semiotic productions. They are done in at least two registers, one of which is the natural language in its oral and/or writing production. Its analysis and interpretation should, therefore, take into account the two sides of the mathematical thinking: we cannot focus on a single register and privilege it as more representative than the others, because *the understanding is at the coordination level of at least two registers.* And yet, this is what is done in all studies based on the "verbal interactions", i.e. oral productions that have been recorded during classroom sessions. They favor the register of language in the oral production mode and are then faced with methodological impasses. Firstly, in the language register, the meaning is based on the discursive operations performed and not the words. Secondly, in oral production, the meaning of what is said is as much in the voice' intonation as in the words spoken, and the transcription of oral exchanges into a written *corpus* eliminates the intonation. Hence, the following questions. How to break into interpretable units the transcribed *corpus* of verbal expressions? Does verbal interaction between various speakers lead to an evolution or an awareness of all the speakers? And what about the students who say nothing? How to escape the trap of an uncontrollable interpretation because it is based on a *corpus* of data that is methodologically impossible to compare with another *corpus*?

The registers allow identifying all the variables about the cognitive processes that command the mathematical understanding. Taking into account these variables is essential to organize teaching sequences or to make students enter the functioning of a mathematical problem solving and mathematical way to formulate a problem based on data observed in reality. The registers allow indeed determining two kinds of cognitive variables.

The first type is related to the conversion of representations, because it is the fundamental cognitive process explicitly required to choose the register in which the representations of hypotheses can explored or mathematically treated. And it also the cognitive process implicitly required to achieve this treatment, because a second register has to be mobilized mentally to anticipate, manage or compensate for the limits of the register chosen for treatment. It why it is important that students can perform fluently the range of possible conversions that are generated between the registers used in the teached mathematics mobilized mentally to anticipate, manage or compensate for the limits of the work register. This conversion fluency cannot be considered as the natural consequence of mathematics learning. Quite the contrary, without any explicit training of the conversion fluency, there no possible mathematics learning for most students, because it is the first condition to become able to recognize the mathematically relevant units of meaning in the content of any representation. And no hope to make students enter the mathematical applications or justifications if this condition is not met!

The second type is related to the treatments. Each register allows performing processing operations on the representations that are not possible in another register. Each of the operations specific to each of the registers is also a cognitive variable. The ability to initiate, develop and control a mathematical treatment, mainly in the non-algorithmic registers (natural language and geometrical configurations), requires that the students become aware of all the possible intrinsic transformations of any semiotic representation into another, within each of these two registers. These possible semiotically intrinsic transformations underlie the various forms of mathematical knowledge and activity in problem solving: definition, visualization, and heuristic exploration, deducing, proof.

The theory of the registers was developed from this methodology, and not the inverse! More precisely, it was developed using the conversion as an analytical tool of variations of representations systematically made within a register. It is an observation device that allows identifying *the persistent misunderstanding phenomena that are not observed* because they are not directly observable in the spontaneous exchanges in class or the single mathematical interpretations of students' productions. It allowed the development of new activities whose goal was the awareness of semiocognitive operations proper to mathematics. And the results obtained led to new questions and new explorations. As Marlene Alves Dias rightly expressed it, "the theory of registers is its own methodology." It is not a hypothesis or explanation, which can be refuted by... a counter-example!

It is meaningless to want to refute the theory of registers without having performed the research methodology that led to its development. We can even ask ourselves about the seriousness of the work of those who intend to do so. *The theory of registers results from the description of the mathematical activity in its completely apart epistemological situation as transformation of semiotic representations.* This description allows us to analyze the cognitive conditions required to understand mathematics. And it is only from this description that the cognitive functioning of thought required to understand mathematics can be analyzed. This functioning lies in the synergistic coordination of heterogeneous registers. Understanding and ability to understand happen and grow on this level. The analysis of students' productions across the curriculum shows that *such coordination is not natural, and mathematics education does not develop it for most students.* We may certainly reject this modeling of cognitive functioning of thought in mathematics, but this leads to denying the cognitive paradox of mathematics and the obstacle that it constitutes when learning mathematics.

Similarly, requiring us to implement a methodology for organizing research according to the registers leads to the thinking that the methodology would be a general formal protocol, *unrelated to the nature of the problems we study, and to the analysis/interpretation procedures of the data collected.* In educational studies in mathematics, data is not primarily quantitative. Statistical treatments are too often based on an analysis of the students' productions *that is limited to a simple codification of successes and errors observed from the single mathematical point of view*! If we are to study truly the recurring and systemic problems of misunderstanding in the learning of mathematics, there is no other choice than the cognitive modelling of mathematical activity in terms of registers to organize the conditions in which the productions of the students will be collected and analyzed. But if not that, then we don't need to care about registers of semiotic representations for teaching mathematics to students ages 6 to 16.

Annex

Analysis of an Example of Introduction of the Linear Function Concept in a Textbook for Students Aged 13–14 Years Old[1]

The introduction of linear function concept is *based on the verbal description of a proportionality table and the equation exemplified by a Cartesian graph.* Both descriptions are mixed and aimed at introducing the mathematical terms. We found on the same page four kinds of juxtaposed representations.

We reorganized them in the table below using the classification of registers shown in Fig. 4.6. Two things are immediately observed. The first is that the cognitive relationship between a proportionality table and a linear function graph implicitly assumes the mobilization of the *geometric representations that we have had to add because it is given with the others.* The second is the set of *register couples* that this introduction mobilizes implicitly. We mark them with arrows. Specifically, it shows *the correspondences between the meaning units of the representation contents that need to be realized to understand.* They are of course left to the student. And the introduction is made as if all the conversions from a register to another were spontaneously obvious. And here there are quite different conversions, without counting the auxiliary representation. What to do if the students do not distinguish nor see any of them? And it is the case for most of them....

[1] *Maths 4ème*. IREM. Strasbourg: Istra, 1992, p. 134.

R. Duval, *Understanding the Mathematical Way of Thinking – The Registers of Semiotic Representations*, DOI 10.1007/978-3-319-56910-9

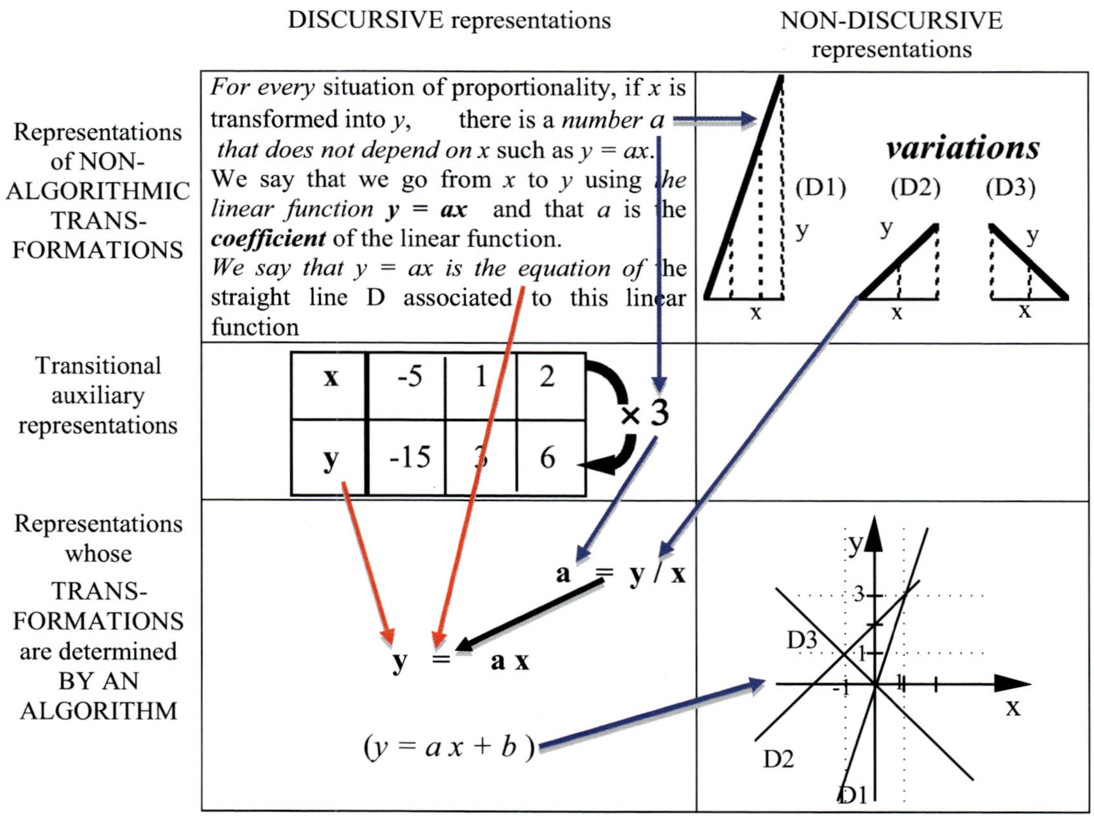
DISCURSIVE representations
NON-DISCURSIVE representations
Representations of NON-ALGORITHMIC TRANS-FORMATIONS
For every situation of proportionality, if x is transformed into y, there is a number a that does not depend on x such as y = ax.
We say that we go from x to y using the linear function y = ax and that a is the coefficient of the linear function.
We say that y = ax is the equation of the straight line D associated to this linear function
variations
(D1) (D2) (D3)
y
x
Transitional auxiliary representations
x -5 1 2
y -15 3 6
× 3
Representations whose TRANS-FORMATIONS are determined BY AN ALGORITHM
a = y / x
y = a x
(y = a x + b)
y
x
3
1
-1
D1
D2
D3

Index

R. Duval, *Understanding the Mathematical Way of Thinking – The Registers of Semiotic Representations*, DOI 10.1007/978-3-319-56910-9

U

V